Performance
Evaluation
and
Benchmarking

Performance Evaluation and Benchmarking

Edited by
Lizy Kurian John
Lieven Eeckhout

Taylor & Francis
Taylor & Francis Group
Boca Raton   London   New York

A CRC title, part of the Taylor & Francis imprint, a member of the
Taylor & Francis Group, the academic division of T&F Informa plc.

Published in 2006 by
CRC Press
Taylor & Francis Group
6000 Broken Sound Parkway NW, Suite 300
Boca Raton, FL 33487-2742

International Standard Book Number-10: 0-8493-3622-8 (Hardcover)
International Standard Book Number-13: 978-0-8493-3622-5 (Hardcover)
Library of Congress Card Number 2005047021

This book contains information obtained from authentic and highly regarded sources. Reprinted material is quoted with permission, and sources are indicated. A wide variety of references are listed. Reasonable efforts have been made to publish reliable data and information, but the author and the publisher cannot assume responsibility for the validity of all materials or for the consequences of their use.

Library of Congress Cataloging-in-Publication Data

John, Lizy Kurian.
    Performance evaluation and benchmarking / Lizy Kurian John and Lieven Eeckhout.
        p. cm.
    Includes bibliographical references and index.
    ISBN 0-8493-3622-8 (alk. paper)
    1. Electronic digital computers--Evaluation. I. Eeckhout, Lieven. II. Title.

QA76.9.E94J64 2005
004.2'4--dc22
                                             2005047021

Taylor & Francis Group
is the Academic Division of T&F Informa plc.

Visit the Taylor & Francis Web site at
http://www.taylorandfrancis.com

and the CRC Press Web site at
http://www.crcpress.com

Preface

It is a real pleasure and honor for us to present you this book titled *Performance Evaluation and Benchmarking*. Performance evaluation and benchmarking is at the heart of computer architecture research and development. Without a deep understanding of benchmarks' behavior on a microprocessor and without efficient and accurate performance evaluation techniques, it is impossible to design next-generation microprocessors. Because this research field is growing and has gained interest and importance over the last few years, we thought it would be appropriate to collect a number of these important recent advances in the field into a research book. This book deals with a large variety of state-of-the-art performance evaluation and benchmarking techniques. The subjects in this book range from simulation models to real hardware performance evaluation, from analytical modeling to fast simulation techniques and detailed simulation models, from single-number performance measurements to the use of statistics for dealing with large data sets, from existing benchmark suites to the conception of representative benchmark suites, from program analysis and workload characterization to its impact on performance evaluation, and other interesting topics. We expect it to be useful to graduate students in computer architecture and to computer architects and designers in the industry.

This book was not entirely written by us. We invited several leading experts in the field to write a chapter on their recent research efforts in the field of performance evaluation and benchmarking. We would like to thank Prof. David J. Lilja from the University of Minnesota, Prof. Tom Conte from North Carolina State University, Prof. Brad Calder from the University of California San Diego, Prof. Chita Das from Penn State, Prof. Brinkley Sprunt from Bucknell University, Alex Mericas from IBM, and Dr. Kishore Menezes from Intel Corporation for accepting our invitation. We thank them and their co-authors for contributing. Special thanks to Dr. Joshua J. Yi from Freescale Semiconductor Inc., Paul D. Bryan from North Carolina State University, Erez Perelman from the University of California San Diego, Prof. Timothy Sherwood from the University of California at Santa Barbara, Prof. Greg Hamerly from Baylor University, Prof. Eun Jung Kim from Texas A&M University, Prof. Ki HwanYum from the University of Texas at San Antonio, Dr. Rumi Zahir from Intel Corporation, and Dr. Susith Fernando from Intel

Corporation for contributing. Many authors went beyond their call to adjust their chapters according to the other chapters. Without their hard work, it would have been impossible to create this book.

We hope you will enjoy reading this book.

Prof. L. K. John
The University of Texas at Austin, USA

Dr. L. Eeckhout
Ghent University, Belgium

Editors

Lizy Kurian John is an associate professor and Engineering Foundation Centennial Teaching Fellow in the electrical and computer engineering department at the University of Texas at Austin. She received her Ph.D. in computer engineering from Pennsylvania State University in 1993. She joined the faculty at the University of Texas at Austin in fall 1996. She was on the faculty at University of South Florida, from 1993 to 1996. Her current research interests are computer architecture, high-performance microprocessors and computer systems, high-performance memory systems, workload characterization, performance evaluation, compiler optimization techniques, reconfigurable computer architectures, and similar topics. She has received several awards including the 2004 Texas Exes teaching award, the 2001 UT Austin Engineering Foundation Faculty award, the 1999 Halliburton Young Faculty award, and the NSF CAREER award. She is a member of IEEE, IEEE Computer Society, ACM, and ACM SIGARCH. She is also a member of Eta Kappa Nu, Tau Beta Pi, and Phi Kappa Phi Honor Societies.

Lieven Eeckhout obtained his master's and Ph.D degrees in computer science and engineering from Ghent University in Belgium in 1998 and 2002, respectively. He is currently working as a postdoctoral researcher at the same university through a grant from the Fund for Scientific Research—Flanders (FWO Vlaanderen). His research interests include computer architecture, performance evaluation, and workload characterization.

Contributors

Paul D. Bryan is a research assistant in the TINKER group, Center for Embedded Systems Research, North Carolina State University. He received his B.S. and M.S. degrees in computer engineering from North Carolina State University in 2002 and 2003, respectively. In addition to his academic work, he also worked as an engineer in the IBM PowerPC Embedded Processor Solutions group from 1999 to 2003.

Brad Calder is a professor of computer science and engineering at the University of California at San Diego. He co-founded the International Symposium on Code Generation and Optimization (CGO) and the ACM Transactions on Architecture and Code Optimization (TACO). Brad Calder received his Ph.D. in computer science from the University of Colorado at Boulder in 1995. He obtained a B.S. in computer science and a B.S. in mathematics from the University of Washington in 1991. He is a recipient of an NSF CAREER Award.

Thomas M. Conte is professor of electrical and computer engineering and director for the Center for Embedded Systems Research at North Carolina State University. He received his M.S. and Ph.D. degrees in electrical engineering from the University of Illinois at Urbana-Champaign in 1988 and 1992, respectively. In addition to academia, he's consulted for numerous companies, including AT&T, IBM, SGI, and Qualcomm, and spent some time in industry as the chief microarchitect of DSP vendor BOPS, Inc. Conte is chair of the IEEE Computer Society Technical Committee on Microprogramming and Microarchitecture (TC-uARCH) as well as a fellow of the IEEE.

Chita R. Das received the M.Sc. degree in electrical engineering from the Regional Engineering College, Rourkela, India, in 1981, and the Ph.D. degree in computer science from the Center for Advanced Computer Studies at the University of Louisiana at Lafayette in 1986. Since 1986, he has been working at Pennsylvania State University, where he is currently a professor in the Department of Computer Science and Engineering. His main areas of interest are parallel and distributed computer architectures, cluster systems, communication networks, resource management in parallel systems, mobile computing, performance evaluation, and fault-tolerant computing. He has published extensively in these areas in all major international journals and conference proceedings. He was an editor of the *IEEE Transactions on Parallel and Distributed Systems* and is currently serving as an editor of the IEEE Transactions on Computers. Dr. Das is a Fellow of the IEEE and is a member of the ACM and the IEEE Computer Society.

Susith Fernando received his bachelor of science degree from the University of Moratuwa in Sri Lanka in 1983. He received the master of science and Ph.D. degrees in computer engineering from Texas A&M University in 1987 and 1994, respectively. Susith joined Intel Corporation in 1996 and has since worked on the Pentium and Itanium projects. His interests include performance monitoring, design for test, and computer architecture.

Greg Hamerly is an assistant professor in the Department of Computer Science at Baylor University. His research area is machine learning and its applications. He earned his M.S. (2001) and Ph.D. (2003) in computer science from the University of California, San Diego, and his B.S. (1999) in computer science from California Polytechnic State University, San Luis Obispo.

Eun Jung Kim received a B.S. degree in computer science from Korea Advanced Institute of Science and Technology in Korea in 1989, an M.S. degree in computer science from Pohang University of Science and Technology in Korea in 1994, and a Ph.D. degree in computer science and engineering from Pennsylvania State University in 2003. From 1994 to 1997, she worked as a member of Technical Staff in Korea Telecom Research and Development Group. Dr. Kim is currently an assistant professor in the Department of Computer Science at Texas A&M University. Her research interests include computer architecture, parallel/distributed systems, computer networks, cluster computing, QoS support in cluster networks and Internet, performance evaluation, and fault-tolerant computing. She is a member of the IEEE Computer Society and of the ACM.

David J. Lilja received Ph.D. and M.S. degrees, both in electrical engineering, from the University of Illinois at Urbana-Champaign, and a B.S. in computer engineering from Iowa State University at Ames. He is currently a professor of electrical and computer engineering at the University of Minnesota in Minneapolis. He has been a visiting senior engineer in the hardware performance analysis group at IBM in Rochester, Minnesota, and a visiting professor at the University of Western Australia in Perth. Previously, he worked as a development engineer at Tandem Computer Incorporated (now a division of Hewlett-Packard) in Cupertino, California. His primary research interests are high-performance computer architecture, parallel computing, hardware-software interactions, nano-computing, and performance analysis.

Kishore Menezes received his bachelor of engineering degree in electronics from the University of Bombay in 1992. He received his master of science degree in computer engineering from the University of South Carolina and a Ph.D. in computer engineering from North Carolina State University. Kishore has worked for Intel Corporation since 1997. While at Intel, Kishore has worked on performance analysis and compiler optimizations. More recently Kishore has been working on implementing architectural enhancements in Itanium firmware. His interests include computer architecture, compilers, and performance analysis.

Alex Mericas obtained his M.S. degree in computer engineering from the National Technological University. He was a member of the POWER4, POWER5, and PPC970 design team responsible for the Hardware Performance Instrumentation. He also led the early performance measurement and verification effort on the POWER4 microprocessor. He currently is a senior technical staff member at IBM in the systems performance area.

Erez Perelman is a senior Ph.D. student at the University of California at San Diego. His research areas include processor architecture and phase analysis. He earned his B.S. (in 2001) in computer science from the University of California at San Diego.

Tim Sherwood is an assistant professor in computer science at the University of California at Santa Barbara. Before joining UCSB in 2003, he received his B.S in computer engineering from UC Davis. His M.S. and Ph.D. are from the University of California at San Diego, where he worked with Professor Brad Calder. His research interests include network and security processors, program phase analysis, embedded systems, and hardware support for software design.

Brinkley Sprunt is an assistant professor of electrical engineering at Bucknell University. Prior to joining Bucknell in 1999, he was a computer architect at Intel for 9 years doing performance projection, analysis, and validation for the 80960CF, Pentium Pro, and Pentium 4 microprocessor design projects. While at Intel, he also developed the hardware performance monitoring architecture for the Pentium 4 processor. His current research interests include computer performance modeling, measurement, and optimization. He developed and maintains the brink and abyss tools that provide a high-level interface to the performance-monitoring capabilities of the Pentium 4 on Linux systems. Sprunt received his M.S. and Ph.D. in electrical and computer engineering from Carnegie Mellon University and his B.S. in electrical engineering from Rice University.

Joshua J. Yi is a recent Ph.D. graduate from the Department of Electrical and Computer Engineering at the University of Minnesota. His Ph.D. thesis research focused on nonspeculative processor optimizations and improving simulation methodology. His research interests include high-performance computer architecture, simulation, and performance analysis. He is currently a performance analyst at Freescale Semiconductor.

Ki Hwan Yum received a B.S. degree in mathematics from Seoul National University in Korea in 1989, an M.S. degree in computer science and engineering from Pohang University of Science and Technology in Korea in 1994, and a Ph.D. degree in computer science and engineering from Pennsylvania State University in 2002. From 1994 to 1997 he was a member of Technical Staff in Korea Telecom Research and Development Group. Dr. Yum is currently an assistant professor in the Department of Computer Science in the University of Texas at San Antonio. His research interests include computer architecture, parallel/distributed systems, cluster computing, and performance evaluation. He is a member of the IEEE Computer Society and of the ACM.

Rumi Zahir is currently a principal engineer at Intel Corporation, where he works on microprocessor and network I/O architectures. Rumi joined Intel in 1992 and was one of the architects responsible for defining the Itanium privileged instruction set, multiprocessing memory model, and performance-monitoring architecture. He applied his expertise in computer architecture and system software to the first-time operating system bring-up efforts on the Merced processor and was one of the main authors of the Itanium programmer's reference manual. Rumi Zahir holds master of science degrees in electrical engineering and computer science and earned his Ph.D. in electrical engineering from the Swiss Federal Institute of Technology in 1991.

Contents

Chapter One

Introduction and Overview

Lizy Kurian John and Lieven Eeckhout

State-of-the-art, high-performance microprocessors contain hundreds of millions of transistors and operate at frequencies close to 4 gigahertz (GHz). These processors are deeply pipelined, execute instructions in out-of-order, issue multiple instructions per cycle, employ significant amounts of speculation, and embrace large on-chip caches. In short, contemporary microprocessors are true marvels of engineering. Designing and evaluating these microprocessors are major challenges especially considering the fact that 1 second of program execution on these processors involves several billions of instructions, and analyzing 1 second of execution may involve dealing with hundreds of billions of pieces of information. The large number of potential designs and the constantly evolving nature of workloads have resulted in performance evaluation becoming an overwhelming task.

Performance evaluation has become particularly overwhelming in early design tradeoff analysis. Several design decisions are made based on performance models before any prototyping is done. Usually, early design analysis is accomplished by simulation models, because building hardware prototypes of state-of-the-art microprocessors is expensive and time consuming. However, simulators are orders of magnitude slower than real hardware. Also, simulation results are artificially *sanitized* in that several unrealistic assumptions might have gone into the simulator. Performance measurements with a prototype will be more accurate; however, a prototype needs to be available. Performance measurement is also valuable after the actual product is available in order to understand the performance of the actual system under various real-world workloads and to identify modifications that could be incorporated in future designs.

This book presents various topics in microprocessor and computer performance evaluation. An overview of modern performance evaluation techniques is presented in Chapter 2. This chapter presents a brief look at prominent methods of performance estimation and measurement. Various simulation methods and hardware performance-monitoring techniques are described as well as their applicability, depending on the goals one wants to achieve.

Benchmarks to be used for performance evaluation have always been controversial. It is extremely difficult to define and identify representative benchmarks. There has been a lot of change in benchmark creation since 1988. In the early days, performance was estimated by the execution latency of a single instruction. Because different instruction types had different execution latencies, the instruction mix was sufficient for accurate performance analysis. Later on, performance evaluation was done largely with small benchmarks such as kernels extracted from applications (e.g., Lawrence Livermore Loops), Dhrystone and Whetstone benchmarks, Linpack, Sort, Sieve of Eratosthenes, 8-Queens problem, Tower of Hanoi, and so forth. The Standard Performance Evaluation Cooperative (SPEC) consortium and the Transactions Processing Council (TPC) formed in 1988 have made available several benchmark suites and benchmarking guidelines. Most of the recent benchmarks have been based on real-world applications. Several state-of-the-art benchmark suites are described in Chapter 3. These benchmark suites reflect different types of workload behavior: general-purpose workloads, Java workloads, database workloads, server workloads, multimedia workloads, embedded workload, and so on.

Another major issue in performance evaluation is the issue of reporting performance with a single number. A single number is easy to understand and easy to be used by the trade press as well as during research and development for comparing design alternatives. The use of multiple benchmarks for performance analysis also makes it necessary to find some kind of an average. The arithmetic mean, geometric mean, and harmonic mean are three ways of finding the central tendency of a group of numbers; however, it should be noted that each of these means should be used under appropriate circumstances. For example, the arithmetic mean can be used to find average execution time from a set of execution times; the harmonic mean can be used to find the central tendency of measures that are in the form of a rate, for example, throughput. However, prior research is not definitive on what means are appropriate for different performance metrics that computer architects use. As a consequence, researchers often use inappropriate mean values when presenting their results. Chapter 4 presents appropriate means to use for various common metrics used while designing and evaluating microprocessors.

Irrespective of whether real system measurement or simulation-based modeling is done, computer architects should use statistical methods to make correct conclusions. For real-system measurements, statistics are useful to deal with noisy data. The noisy data comes from noise in the system being measured or is due to the measurement tools themselves. For simulation-based modeling the major challenge is to deal with huge amounts of data and to observe trends in the data. For example, at processor design time, a large number of microarchitectural design parameters need to be fine-tuned. In addition, complex interactions between these microarchitectural parameters complicate the design space exploration process even further. The end result is that in order to fully understand the complex interaction of a computer program's execution with the underlying microprocessor, a huge number of simulations are required. Statistics can be really helpful for simulation-based design studies to cut down the number of simulations that need to be done without compromising the end result. Chapter 5 describes several statistical techniques to rigorously guide performance analysis.

To date, the de facto standard for early stage performance analysis is detailed processor simulation using real-life benchmarks. An important disadvantage of this approach is that it is prohibitively time consuming. The main reason is the large number of instructions that need to be simulated per benchmark. Nowadays, it is not exceptional that a benchmark has a dynamic instruction count of several hundreds of billions of instructions. Simulating such huge instruction counts can take weeks for completion even on today's fastest machines. Therefore researchers have proposed several techniques for speeding up these time-consuming simulations. These approaches are discussed in Chapters 6, 7 and 8.

Random sampling or the random selection of instruction intervals throughout the entire benchmark execution is one approach for reducing the total simulation time. Instead of simulating the entire benchmark only the samples are to be simulated. By doing so, significant simulation speedups can be obtained while attaining highly accurate performance estimates. There is, however, one issue that needs to be dealt with—the unknown hardware state at the beginning of each sample during sampled simulation. To address that problem, researchers have proposed functional warming prior to each sample. Random sampling and warm-up techniques are discussed in Chapter 6.

Chapter 7 presents SimPoint, which is an intelligent sampling approach that selects samples called simulation points (in SimPoint terminology), based on a program's phase behavior. Instead of randomly selecting samples, SimPoint first determines the large-scale phase behavior of a program execution and subsequently picks one simulation point from each phase of execution.

A radically different approach to sampling is statistical simulation. The idea of statistical simulation is to collect a number of important program execution characteristics and generate a synthetic trace from it. Because of the statistical nature of this technique, simulation of the synthetic trace quickly converges to a steady-state value. As such, a very short synthetic trace suffices to attain a performance estimates. Chapter 8 describes statistical simulation as a viable tool for efficient early design stage explorations.

In contemporary research and development, multiple benchmarks with multiple input data sets are simulated from multiple benchmark suites. However, there exists significant redundancy across inputs and across programs. Chapter 9 describes methods to identify such redundancy in benchmarks so that only relevant and distinct benchmarks need to be simulated.

Although quantitative evaluation has been popular in the computer architecture field, there are several cases for which analytical modeling can be used. Chapter 10 introduces the fundamentals of analytical modeling.

Chapters 11, 12, and 13 describe performance-monitoring facilities on three state-of-the-art microprocessors. Such measurement infrastructure is available on all modern day high-performance processors to make it easy to obtain information of actual performance on real hardware. These chapters discuss the performance monitoring abilities of Intel Pentium, IBM POWER, and Intel Itanium processors.

Chapter Two

Performance Modeling and Measurement Techniques

Lizy Kurian John

Contents

Performance evaluation can be classified into performance modeling and performance measurement. Performance modeling is typically used in early stages of the design process, when actual systems are not available for measurement or if the actual systems do not have test points to measure every detail of interest. Performance modeling may further be divided into

Table 2.1 A Classification of Performance Evaluation Techniques

Performance Modeling	Simulation	Trace-Driven Simulation Execution-Driven Simulation Complete System Simulation Event-Driven Simulation Statistical Simulation Probabilistic Models
	Analytical Modeling	Queuing Models Markov Models Petri Net Models
Performance Measurement	On-Chip Hardware Monitoring (e.g., Performance-monitoring counters) Off-Chip Hardware Monitoring Software Monitoring Microcoded Instrumentation	

simulation-based modeling and analytical modeling. Simulation models may further be classified into numerous categories depending on the mode or level of detail. Analytical models use mathematical principles to create probabilistic models, queuing models, Markov models, or Petri nets. Performance modeling is inevitable during the early design phases in order to understand design tradeoffs and arrive at a good design. Measuring actual performance is certainly likely to be more accurate; however, performance measurement is possible only if the system of interest is available for measurement and only if one has access to the parameters of interest. Performance measurement on the actual product helps to validate the models used in the design process and provides additional feedback for future designs. One of the drawbacks of performance measurement is that performance of only the existing configuration can be measured. The configuration of the system under measurement often cannot be altered, or, in the best cases, it might allow limited reconfiguration. Performance measurement may further be classified into on-chip hardware monitoring, off-chip hardware monitoring, software monitoring, and microcoded instrumentation. Table 2.1 illustrates a classification of performance evaluation techniques.

There are several desirable features that performance modeling/measurement techniques and tools should possess:

- They must be accurate. Because performance results influence important design and purchase decisions, accuracy is important. It is easy to build models/techniques that are heavily sanitized; however, such models will not be accurate.
- They must not be expensive. Building the performance evaluation or measurement facility should not cost a significant amount of time or money.

- They must be easy to change or extend. Microprocessors and computer systems constantly undergo changes, and it must be easy to extend the modeling/measurement facility to the upgraded system.
- They must not need the source code of applications. If tools and techniques necessitate source code, it will not be possible to evaluate commercial applications where source is not often available.
- They should measure all activity, including operating system and user activity. It is often easy to build tools that measure only user activity. This was acceptable in traditional scientific and engineering workloads; however, database, Web server, and Java workloads have significant operating system activity, and it is important to build tools that measure operating system activity as well.
- They should be capable of measuring a wide variety of applications, including those that use signals, exceptions, and DLLs (Dynamically Linked Libraries).
- They should be user-friendly. Hard-to-use tools are often underutilized and may also result in more user error.
- They must be noninvasive. The measurement process must not alter the system or degrade the system's performance.
- They should be fast. If a performance model is very slow, long-running workloads that take hours to run may take days or weeks to run on the model. If evaluation takes weeks and months, the extent of design space exploration that can be performed will be very limited. If an instrumentation tool is slow, it can also be invasive.
- Models should provide control over aspects that are measured. It should be possible to selectively measure what is required.
- Models and tools should handle multiprocessor systems and multithreaded applications. Dual- and quad-processor systems are very common nowadays. Applications are becoming increasingly multithreaded, especially with the advent of Java, and it is important that the tool handles these.
- It will be desirable for a performance evaluation technique to be able to evaluate the performance of systems that are not yet built.

Many of these requirements are often conflicting. For instance, it is difficult for a mechanism to be fast and accurate. Consider mathematical models. They are fast; however, several simplifying assumptions go into their creation and often they are not accurate. Similarly, many users like graphical user interfaces (GUIs), which increase the user-friendly nature, but most instrumentation and simulation tools with GUIs are slow and invasive.

2.1   *Performance modeling*

Performance measurement can be done only if the actual system or a prototype exists. It is expensive to build prototypes for early design-stage evaluation. Hence one would need to resort to some kind of modeling in order to study systems yet to be built. Performance modeling can be done using simulation models or analytical models.

2.1.1   Simulation

Simulation has become the de facto performance-modeling method in the evaluation of microprocessor and computer architectures. There are several reasons for this. The accuracy of analytical models in the past has been insufficient for the type of design decisions that computer architects wish to make (for instance, what kind of caches or branch predictors are needed, or what kind of instruction windows are required). Hence, cycle accurate simulation has been used extensively by computer architects. Simulators model existing or future machines or microprocessors. They are essentially a model of the system being simulated, written in a high-level computer language such as C or Java, and running on some existing machine. The machine on which the simulator runs is called the host machine, and the machine being modeled is called the target machine. Such simulators can be constructed in many ways.

Simulators can be functional simulators or timing simulators. They can be trace-driven or execution-driven simulators. They can be simulators of components of the system or that of the complete system. Functional simulators simulate the functionality of the target processor and, in essence, provide a component similar to the one being modeled. The register values of the simulated machine are available in the equivalent registers of the simulator. Pure functional simulators only implement the functionality and merely help to validate the correctness of an architecture; however, they can be augmented to include performance information. For instance, in addition to the values, the simulators can provide performance information in terms of cycles of execution, cache hit ratios, branch prediction rates, and so on. Such a simulator is a virtual component representing the microprocessor or subsystem being modeled plus a variety of performance information.

If performance evaluation is the only objective, functionality does not need to be modeled. For instance, a cache performance simulator does not need to actually store values in the cache; it only needs to store information related to the address of the value being cached. That information is sufficient to determine a future hit or miss. Operand values are not necessary in many performance evaluations. However, if a technique such as value prediction is being evaluated, it would be important to have the values. Although it is nice to have the values as well, a simulator that models functionality in addition to performance is bound to be slower than a pure performance simulator.

2.1.1.1   Trace-driven simulation

Trace-driven simulation consists of a simulator model whose input is modeled as a trace or sequence of information representing the instruction sequence that would have actually executed on the target machine. A simple trace-driven cache simulator needs a trace consisting of address values. Depending on whether the simulator is modeling an instruction, data, or a unified cache, the address trace should contain addresses of instruction and data references.

Cachesim5 [1] and Dinero IV [2] are examples of cache simulators for memory reference traces. Cachesim5 comes from Sun Microsystems along with their SHADE package [1]. Dinero IV [2] is available from the University of Wisconsin at Madison. These simulators are not timing simulators. There is no notion of simulated time or cycles; information is only about memory references. They are not functional simulators. Data and instructions do not move in and out of the caches. The primary result of simulation is hit and miss information. The basic idea is to simulate a memory hierarchy consisting of various caches. The different parameters of each cache can be set separately (architecture, mapping policies, replacement policies, write policy, measured statistics). During initialization, the configuration to be simulated is built up, one cache at a time, starting with each memory as a special case. After initialization, each reference is fed to the appropriate top-level cache by a single, simple function call. Lower levels of the hierarchy are handled automatically. Trace-driven simulation does not necessarily mean that a trace is stored. One can have a tracer/profiler to feed the trace to the simulator on-the-fly so that the trace storage requirements can be eliminated. This can be done using a Unix pipe or by creating explicit data structures to buffer blocks of trace. If traces are stored and transferred to simulation environments, typically trace compression techniques are used to reduce storage requirements [3–4].

Trace-driven simulation can be used not only for caches, but also for entire processor pipelines. A trace for a processor simulator should contain information on instruction opcodes, registers, branch offsets, and so on.

Trace-driven simulators are simple and easy to understand. They are easy to debug. Traces can be shared to other researchers/designers and repeatable experiments can be conducted. However, trace-driven simulation has two major problems:

1. Traces can be prohibitively long if entire executions of some real-world applications are considered. Trace size is proportional to the dynamic instruction count of the benchmark.
2. The traces are not very representative inputs for modern out-of-order processors. Most trace generators generate traces of only completed or retired instructions in speculative processors. Hence they do not contain instructions from the mispredicted path.

The first problem is typically solved using trace sampling and trace reduction techniques. Trace sampling is a method to achieve reduced traces. However, the sampling should be performed in such a way that the resulting trace is representative of the original trace. It may not be sufficient to periodically sample a program execution. Locality properties of the resulting sequence may be widely different from that of the original sequence. Another technique is to skip tracing for a certain interval, collect for a fixed interval, and then skip again. It may also be needed to leave a warm-up period after the skip interval, to let the caches and other such structures warm up [5]. Several trace sampling techniques are discussed by Crowley and Baer [6–8]. The QPT trace collection system [9] solves the trace size issue by splitting the tracing process into a trace record generation step and a trace regeneration process. The trace record has a size similar to the static code size, and the trace regeneration expands it to the actual, full trace upon demand.

The second problem can be solved by reconstructing the mispredicted path [10]. An image of the instruction memory space of the application is created by one pass through the trace, and thereafter fetching from this image as opposed to the trace. Although 100% of the mispredicted branch targets may not be in the recreated image, studies show that more than 95% of the targets can be located. Also, it has been shown that performance inaccuracy due to the absence of mispredicted paths is not very high [11–12].

2.1.1.2   *Execution-driven simulation*

There are two contexts in which terminology for execution-driven simulation is used by researchers and practitioners. Some refer to simulators that take program executables as input as execution-driven simulators. These simulators utilize the actual input executable and not a trace. Hence the size of the input is proportional to the static instruction count and not the dynamic instruction count. Mispredicted paths can be accurately simulated as well. Thus these simulators solve the two major problems faced by trace-driven simulators, namely the storage requirements for large traces and the inability to simulate instructions along mispredicted paths. The widely used SimpleScalar simulator [13] is an example of such an execution-driven simulator. With this tool set, the user can simulate real programs on a range of modern processors and systems, using fast executable-driven simulation. There is a fast functional simulator and a detailed, out-of-order issue processor that supports nonblocking caches, speculative execution, and state-of-the-art branch prediction.

Some others consider execution-driven simulators to be simulators that rely on actual execution of parts of code on the host machine (hardware acceleration by the host instead of simulation) [14]. These execution-driven simulators do not simulate every individual instruction in the application; only the instructions that are of interest are simulated. The remaining instructions are directly executed by the host computer. This can be done when the instruction set of the host is the same as that of the machine being simulated. Such simulation involves two stages. In the first stage, or preprocessing, the application program is modified by inserting calls to the simulator routines at events of interest. For instance, for a memory system simulator, only memory access instructions need to be instrumented. For other instructions, the only important thing is to make sure that they get performed and that their execution time is properly accounted for. The advantage of this type of execution-driven simulation is speed. By directly executing most instructions at the machine's execution rate, the simulator can operate orders of magnitude faster than cycle-by-cycle simulators that emulate each individual instruction. Tango, Proteus, and FAST are examples of such simulators [14].

Execution-driven simulation is highly accurate but is very time consuming and requires long periods of time for developing the simulator.

Creating and maintaining detailed cycle-accurate simulators are difficult software tasks. Processor microarchitectures change very frequently, and it would be desirable to have simulator infrastructures that are reusable, extensible, and easily modifiable. Principles of software engineering can be applied here to create modular simulators. Asim [15], Liberty [16], and MicroLib [17] are examples of execution driven-simulators built with the philosophy of modular components. Such simulators ease the challenge of incorporating modifications.

Detailed execution-driven simulation of modern benchmarks on state-of-the-art architectures take prohibitively long simulation times. As in trace-driven simulation, sampling provides a solution here. Several approaches to perform sampled simulation have been developed. Some of those approaches are described in Chapters 6 and 7 of this book.

Most of the simulators that have been discussed so far are for superscalar microprocessors. Intel IA-64 and several media processors use the VLIW (very long instruction word) architecture. The TRIMARAN infrastructure [18] includes a variety of tools to compile to and estimate performance of VLIW or EPIC-style architectures.

Multiprocessor and multithreaded architectures are becoming very common. Although SimpleScalar can only simulate uniprocessors, derivatives such as MP_simplesim [19] and SimpleMP [20] can simulate multiprocessor caches and multithreaded architectures, respectively. Multiprocessors can also be simulated by using simulators such as Tango, Proteus and FAST [14].

2.1.1.3 Complete system simulation

Many execution- and trace-driven simulators only simulate the processor and memory subsystem. Neither input/output (I/O) activity nor operating system (OS) activity is handled in simulators like SimpleScalar. But in many workloads, it is extremely important to consider I/O and OS activity. Complete system simulators are complete simulation environments that model hardware components with enough detail to boot and run a full-blown commercial OS. The functionality of the processors, memory subsystem, disks, buses, SCSI/IDE/FC controllers, network controllers, graphics controllers, CD-ROM, serial devices, timers, and so on are modeled accurately in order to achieve this. Although functionality stays the same, different microarchitectures in the processing component can lead to different performance. Most of the complete system simulators use microarchitectural models that can be plugged in. For instance, SimOS [21], a popular complete system simulator allows three different processor models, one extremely simple processor, one pipelined, and one aggressive superscalar model. SimOS [21] and SIMICS [22] can simulate uniprocessor and multiprocessor systems. SimOS natively models the MIPS instruction set architecture (ISA), whereas SIMICS models the SPARC ISA. Mambo [23] is another emerging complete system simulator that models the PowerPC ISA. Many of these

simulators can cross-compile and cross-simulate other ISAs and architectures.

The advantage of full-system simulation is that the activity of the entire
system, including operating system, can be analyzed. Ignoring operating
system activity may not have significant performance impact in SPEC-CPU
type of benchmarks; however, database and commercial workloads spend
close to half of their execution in operating system code, and no reasonable
evaluation of their performance can be performed without considering OS
activity. Full-system simulators are very accurate but are extremely slow.
They are also difficult to develop.

2.1.1.4   *Event-driven simulation*

The simulators described in the previous three subsections simulate performance on a cycle-by-cycle basis. In cycle-by-cycle simulation, each cycle of
the processor is simulated. A cycle-by-cycle simulator mimics the operation
of a processor by simulating each action in every cycle, such as fetching,
decoding, and executing. Each part of the simulator performs its job for that
cycle. In many cycles, many units may have no task to perform, but it realizes
that only after it "wakes up" to perform its task. The operation of the
simulator matches our intuition of the working of a processor or computer
system but often produces very slow models.

An alternative is to create a simulator where events are scheduled for
specific times and simulation looks at all the scheduled events and performs
simulation corresponding to the events (as opposed to simulating the processor cycle-by-cycle). In an event-driven simulation, tasks are posted to an
event queue at the end of each simulation cycle. During each simulation cycle,
a scheduler scans the events in the queue and services them in the time-order
in which they are scheduled for. If the current simulation time is 400 cycles
and the earliest event in the queue is to occur at 500 cycles, the simulation
time advances to 500 cycles. Event-driven simulation is used in many fields
other than computer architecture performance evaluation. A very common
example is VHDL simulation. Event-driven and cycle-by-cycle simulation
styles can be combined to create models where parts of a model are simulated
in detail regardless of what is happening in the processor, and other parts
are invoked only when there is an event. Reilly and Edmondson created such
a model for the Alpha microprocessor modeling some units on a
cycle-by-cycle basis while modeling other units on an event-driven basis [24].

When event-driven simulation is applied to computer performance evaluation, the inputs to the simulator can be derived stochastically rather than
as a trace/executable from an actual execution. For instance, one can con-
struct a memory system simulator in which the inputs are assumed to arrive
according to be a Gaussian distribution. Such models can be written in
general-purpose languages such as C, or using special simulation languages
such as SIMSCRIPT. Languages such as SIMSCRIPT have several built-in
primitives to allow quick simulation of most kinds of common systems.
There are built-in input profiles, resource templates, process templates,

queue structures, and so on to facilitate easy simulation of common systems. An example of the use of event-driven simulators using SIMSCRIPT may be seen in the performance evaluation of multiple-bus multiprocessor systems in John et al. [25]. The statistical simulation described in the next subsection statistically creates a different input trace corresponding to each benchmark that one wants to simulate, whereas in the stochastic event-driven simulator, input models are derived more generally. It may also be noticed that a statistically generated input trace can be fed to a trace-driven simulator that is not event-driven.

2.1.1.5   Statistical simulation

Statistical simulation [26–28] is a simulation technique that uses a statistically generated trace along with a simulation model where many components are modeled only statistically. First, benchmark programs are analyzed in detail to find major program characteristics such as instruction mix, cache and branch misprediction rates, and so on. Then, an artificial input sequence with approximately the same program characteristics is statistically generated using random number generators. This input sequence (a synthetic trace) is fed to a simulator that estimates the number of cycles taken for executing each of the instructions in the input sequence. The processor is modeled at a reduced level of detail; for instance, cache accesses may be deemed as hits or misses based on a statistical profile as opposed to actual simulation of a cache. Experiments with such statistical simulations [26] show that IPC of SPECint-95 programs can be estimated very quickly with reasonable accuracy. The statistically generated instructions matched the characteristics of unstructured control flow in SPECint programs easily; however, additional characteristics needed to be modeled in order to make the technique work with programs that have regular control flow. Recent experiments with statistical simulation [27–28] demonstrate that performance estimates on SPEC2000 integer and floating-point programs can be obtained with orders of magnitude more speed than execution-driven simulation. More details on statistical simulation can be found in Chapter 8.

2.1.2   Program profilers

There is a class of tools called software profiling tools, which are similar to simulators and performance measurement tools. These tools are used to profile programs, that is, to obtain instruction mixes, register usage statistics, branch distance distribution statistics, or to generate traces. These tools can also be thought of as software monitoring on a simulator. They often accept program executables as input and decode and analyze each instruction in the executable. These program profilers can also be used as the front end of simulators.

Profiling tools typically add instrumentation code into the original program, inserting code to perform run-time data collection. Some perform the instrumentation during source compilation, whereas most do it either during linking or after the executable is generated. Executable-level instrumentation is harder than source-level instrumentation, but leads to tools that can profile applications whose sources are not accessible (e.g., proprietary software packages).

Several program profiling tools have been built for various ISAs, especially soon after the advent of RISC ISAs. Pixie [29], built for the MIPS ISA was an early instrumentation tool that was very widely used. Pixie performed the instrumentation at executable level and generated an instrumented executable often called the *pixified* program. Other similar tools are nixie for MIPS [30]; SPIX [30] and SHADE for SPARC [1,30]; IDtrace for IA-32 [30]; Goblin for IBM RS 6000 [30]; and ATOM for Alpha [31]. All of these perform executable-level instrumentation. Examples for tools built to perform compile-time instrumentation are AE [32] and Spike [30], which are integrated with C compilers. There is also a new tool called PIN for the IA-32 [33], which performs the instrumentation at run-time as opposed to compile-time or link-time. It should be remembered that profilers are not completely noninvasive; they cause execution-time dilation and use processor registers for the profiling process. Although it is easy to build a simple profiling tool that simply interprets each instruction, many of these tools have incorporated carefully thought-out techniques to improve the speed of the profiling process and to minimize the invasiveness. Many of these profiling tools also incorporate a variety of utilities or hooks to develop custom analysis programs. This chapter will just describe SHADE as an example of executable instrumentation before run-time and PIN as an example of run-time instrumentation.

SHADE: SHADE is a fast instruction-set simulator for execution profiling [1]. It is a simulation and tracing tool that provides features of simulators and tracers in one tool. SHADE analyzes the original program instructions and cross-compiles them to sequences of instructions that simulate or trace the original code. Static cross-compilation can produce fast code, but purely static translators cannot simulate and trace all details of dynamically linked code. If the libraries are already instrumented, it is possible to get profiles from the dynamically linked code as well. One can develop a variety of analyzers to process the information generated by SHADE and create the performance metrics of interest. For instance, one can use SHADE to generate address traces to feed into a cache analyzer to compute hit rates and miss rates of cache configurations. The SHADE analyzer Cachesim5 does exactly this.

PIN [33]: PIN is a relatively new program instrumentation tool that performs the instrumentation at run-time as opposed to compile-time or link-time. PIN supports Linux executables for IA-32 and Itanium processors. PIN does not create an instrumented version of the executable but rather adds the instrumentation code while the executable is running. This makes it possible to attach PIN to an already running process. PIN automatically saves and restores the registers that are overwritten by the injected code. PIN is a versatile tool that includes several utilities such as basic block profilers, cache simulators, and trace generators.

With the advent of Java, virtual machines, and binary translation, pro-filers can be required to profile at multiple levels. Although Java programs can be traced using SHADE or another instruction set profiler to obtain profiles of native execution, one might need profiles at the bytecode level. Jaba [34] is a Java bytecode analyzer developed at the University of Texas for tracing Java programs. It used JVM (Java Virtual Machine) specification 1.1. It allows the user to gather information about the dynamic execution of a Java application at the Java bytecode level. It provides information on byte-codes executed, load operations, branches executed, branch outcomes, and so on. Information about the use of this tool can be found in Radhakrishnan, Rubio, and John [35].

2.1.3   Analytical modeling

Analytical performance models, although not popular for microprocessors, are suitable for the evaluation of large computer systems. In large systems where details cannot be modeled accurately through cycle accurate simula-tion, analytical modeling is an appropriate way to obtain approximate per-formance metrics. Computer systems can generally be considered as a set of hardware and software resources and a set of tasks or jobs competing for using the resources. Multicomputer systems and multiprogrammed systems are examples.

Analytical models rely on probabilistic methods, queuing theory, Markov models, or Petri nets to create a model of the computer system. A large body of literature on analytical models of computers exists from the 1970s and early 1980s. Heidelberger and Lavenberg [36] published an article summarizing research on computer performance evaluation models. This arti-cle contains 205 references, which cover most of the work on performance evaluation until 1984.

Analytical models are cost-effective because they are based on efficient solutions to mathematical equations. However, in order to be able to have tractable solutions, simplifying assumptions are often made regarding the structure of the model. As a result, analytical models do not capture all the details typically built into simulation models. It is generally thought that care-fully constructed analytical models can provide estimates of average job throughputs and device utilizations to within 10% accuracy and average response times within 30% accuracy. This level of accuracy, although insuffi-cient for microarchitectural enhancement studies, is sufficient for capacity plan-ning in multicomputer systems, I/O subsystem performance evaluation in large server farms, and in early design evaluations of multiprocessor systems.

There has not been much work on analytical modeling of microprocessors. The level of accuracy needed in trade-off analysis for microprocessor structures is more than what typical analytical models can provide. However, some effort into this arena came from Noonburg and Shen [37] and Sorin et al [38] and Karkhanis and Smith [39]. Those interested in modeling superscalar processors using analytical models should read these references. Noonburg and Shen used a Markov model to model a pipelined processor. Sorin et al. used probabilistic techniques to model a multiprocessor composed of superscalar processors. Karkhanis and Smith proposed a first-order superscalar processor model that models steady-state performance under ideal conditions and transient performance penalties due to branch mispredictions, instruction cache misses, and data cache misses. Queuing theory is also applicable to superscalar processor modeling, because modern superscalar processors contain instruction queues in which instructions wait to be issued to one among a group of functional units. These analytical models can be very useful in the earliest stages of the microprocessor design process. In addition, these models can reveal interesting insight into the internals of a superscalar processor. Analytical modeling is further explored in Chapter 10 of this book.

The statistical simulation technique described earlier can be considered as a hybrid of simulation and analytical modeling techniques. It, in fact, models the simulator input using a probabilistic model. Some operations of the processor are also modeled probabilistically. Statistical simulation thus has advantages of both simulation and analytical modeling.

2.2   *Performance measurement*

Performance measurement is used for understanding systems that are already built or prototyped. There are several major purposes performance measurement can serve:

- Tune systems that have been built.
- Tune applications if source code and algorithms can still be changed.
- Validate performance models that were built.
- Influence the design of future systems to be built.

Essentially, the process involves

1. Understanding the bottlenecks in systems that have been built.
2. Understanding the applications that are running on the system and the match between the features of the system and the characteristics of the workload, and,
3. Innovating design features that will exploit the workload features.

Performance measurement can be done via the following means:

- On-chip hardware monitoring
- Off-chip hardware monitoring
- Software monitoring
- Microcoded instrumentation

Many systems are built with configurable features. For instance, some microprocessors have control registers (switches) that can be programmed to turn on or off features like branch prediction, prefetching, and so on [40]. Measurement on such processors can reveal very critical information on effectiveness of microarchitectural structures, under real-world workloads. Often, microprocessor companies will incorporate such (undisclosed) switches. It is one way to safeguard against features that could not be conclusively evaluated by performance models.

2.2.1   On-chip performance monitoring counters

All state-of-the-art, high-performance microprocessors, including Intel's Pentium 3 and Pentium 4, IBM's POWER4 and POWER5 processors, AMD's Athlon, Compaq's Alpha, and Sun's UltraSPARC processors, incorporate on-chip performance-monitoring counters that can be used to understand performance of these microprocessors while they run complex, real-world workloads. This ability has overcome a serious limitation of simulators, that they often could not execute complex workloads. Now, complex run-time systems involving multiple software applications can be evaluated and monitored very closely. All microprocessor vendors nowadays release information on their performance-monitoring counters, although they are not part of the architecture.

The performance counters can be used to monitor hundreds of different performance metrics, including cycle count, instruction counts at fetch/ decode/retire, cache misses (at the various levels), and branch mispredictions. The counters are typically configured and accessed with special instructions that access special control registers. The counters can be made to measure user and kernel activity in combination or in isolation. Although hundreds of distinct events can be measured, often only 2 to 10 events can be measured simultaneously. At times, certain events are restricted to be accessible only through a particular counter. These steps are necessary to reduce the hardware overhead associated with on-chip performance monitoring. Performance counters do consume on-chip real estate. Unless carefully implemented, they can also impact the processor cycle time. Out-of-order execution also complicates the hardware support required to conduct on-chip performance measurements [41].

Several studies in the past illustrate how performance-monitoring counters can be used to analyze performance of real-world workloads. Bhandarkar and Ding [42] analyzed Pentium 3 performance counter results to understand the out-of-order execution of Pentium 3 (in comparison) to in-order superscalar execution of Pentium 2. Luo et al [43] investigated the major differences between SPEC CPU workloads and commercial workloads by studying Web server and e-commerce workloads in addition to SPECint2000 programs. Vtune [44], PMON [45] and Brink-Abyss [46] are examples of tools that facilitate performance measurements on modern microprocessors.

Chapters 11, 12, and 13 of this book describe performance-monitoring facilities on three state-of-the-art microprocessors. Similar resources exist

on most modern microprocessors. Chapter 11 is written by the author of
the Brink-Abyss tool. This kind of measurement provides an opportunity
to validate simulation experiments with actual measurements of realistic
workloads on real systems. One can measure user and operating system
activity separately, using these performance monitors. Because everything
on a processor is counted, effort should be made to have minimal or no
other undesired processes running during experimentation. This type of
performance measurement can be done on executables (i.e., no source code
is needed).

2.2.2   Off-chip hardware monitoring

Instrumentation using hardware means can also be done by attaching
off-chip hardware. Two examples from AMD are used to describe this type
of tool.

SpeedTracer from AMD: AMD developed this hardware-tracing platform to aid in the design of their x86 microprocessors. When an application
is being traced, the tracer interrupts the processor on each instruction boundary. The state of the CPU is captured on each interrupt and then transferred
to a separate control machine where the trace is stored. The trace contains
virtually all valuable pieces of information for each instruction that executes
on the processor. Operating system activity can also be traced. However,
tracing in this manner can be invasive and may slow down the processor.
Although the processor is running slower, external events such as disk and
memory accesses still happen in real time, thus looking very fast to the
slowed-down processor. Usually this issue is addressed by adjusting the
timer interrupt frequency. Use of this performance-monitoring facility can
be seen in Merten et al. [47] and Bhargava et al. [48].

Logic Analyzers: Poursepanj and Christie [49] use a Tektronix TLA 700
logic analyzer to analyze 3-D graphics workloads on AMD-K6-2–based systems. Detailed logic analyzer traces are limited by restrictions on sizes and
are typically used for the most important sections of the program under
analysis. Preliminary coarse-level analysis can be done by performance-monitoring counters and software instrumentation. Poursepanj and Christie used
logic analyzer traces for a few tens of frames that included a second or two
of smooth motion [49].

2.2.3   Software monitoring

Software monitoring is often performed by utilizing architectural features
such as a trap instruction or a breakpoint instruction on an actual system,
or on a prototype. The VAX processor from Digital (now Compaq) had a
T-bit that caused an exception after every instruction. Software monitoring
used to be an important mode of performance evaluation before the advent
of on-chip performance-monitoring counters. The primary advantage of software monitoring is that it is easy to do. However, disadvantages include

that the instrumentation can slow down the application. The overhead of servicing the exception, switching to a data collection process, and performing the necessary tracing can slow down a program by more than 1000 times. Another disadvantage is that software-monitoring systems typically handle only the user activity. It is extremely difficult to create a software-monitoring system that can monitor operating system activity.

2.2.4 Microcoded instrumentation

Digital (now Compaq) used microcoded instrumentation to obtain traces of VAX and Alpha architectures. The ATUM tool [50] used extensively by Digital in the late 1980s and early 1990s used microcoded instrumentation. This was a technique lying between trapping information on each instruction using hardware interrupts (traps) and software traps. The tracing system essentially modified the VAX microcode to record all instruction and data references in a reserved portion of memory. Unlike software monitoring, ATUM could trace all processes, including the operating system. However, this kind of tracing is invasive and can slow down the system by a factor of 10 without including the time to write the trace to the disk.

One difference between modern on-chip hardware monitoring and microcoded instrumentation is that, typically, this type of instrumentation recorded the instruction stream but not the performance.

2.3 Energy and power simulators

Power dissipation and energy consumption have become important design constraints in addition to performance. Hence it has become important for computer architects to evaluate their architectures from the perspective of power dissipation and energy consumption. Power consumption of chips comes from activity-based dynamic power or activity-independent static power. The first step in estimating dynamic power consumption is to build power models for individual components inside the processor microarchitecture. For instance, models should be built to reflect the power associated with processor functional units, register read and write accesses, cache accesses, reorder buffer accesses, buses, and so on. Once these models are built, dynamic power can be estimated based on the activity in each unit. Detailed cycle-accurate performance simulators contain the information on activity of the various components and, hence, energy consumption estimation can be integrated to performance estimation. Wattch [51] is such a simulator that incorporates power models into the popular SimpleScalar performance simulator. The SoftWatt [52] simulator incorporates power models to the SimOS complete system simulator. POWER-IMPACT [53] incorporates power models to the IMPACT VLIW performance simulator environment. If cache power needs to be modeled in detail, the CACTI tool [54] can be used, which models power, area, and timing. CACTI has models for various cache mapping schemes, cache array layouts, and port configurations.

Power consumption of chips used to be dominated by activity-based dynamic power consumption; however, with shrinking feature sizes, leakage power is becoming a major component of the chip power consumption. HotLeakage [55] includes software models to estimate leakage power considering supply voltage, gate leakage, temperature, and other factors. Parameters derived from circuit-level simulation are used to build models for building blocks, which are integrated to make models for components inside modern microprocessors. The tool can model leakage in a variety of structures, including caches. The tool can be integrated with simulators such as Wattch.

2.4   Validation

It is extremely important to validate performance models and measurements. Many performance models are often heavily sanitized. Operating system and other real-world effects can make measured performance very different from simulated performance. Models can be validated by measurements on actual systems. Measurements are not error-free either. Any measurements dealing with several variables are prone to human error during usage. Simulations and measurements must be validated with small input sequences where outcome can be predicted without complex models. Approximate estimates calculated using simple heuristic models or analytical models should be used to validate simulation models. It should always be remembered that higher precision (or increased number of decimal places) does not substitute accuracy. Confidence in simulators and measurement facilities should be built with systematic performance validations. Examples of this process can be seen in [56][57][58].

2.5   Conclusion

There are a variety of ways in which performance can be estimated and measured. They vary in the level of detail modeled, complexity, accuracy, and development time. Different models are appropriate under different situations. Appropriate models should be used depending on the specific purpose of the evaluation. Detailed cycle-accurate simulation is not called for in many design decisions. One should always check the sanity of the assumptions that have gone into creation of detailed models and evaluate whether they are applicable for the specific situation being evaluated at the moment. Rather than trusting numbers spit out by detailed simulators as golden values, simple sanity checks and validation exercises should be frequently done.

This chapter does not do a comprehensive treatment of any of the simulation methodologies but has given to the reader some pointers for further study, research, and development. The resources listed at the end of the chapter provide more detailed explanations. The computer architecture home page [59] also provides information on tools for architecture research and performance evaluation.

References

1. Cmelik, B. and Keppel, D., SHADE: A fast instruction-set simulator for execution profiling, in *Fast Simulation of Computer Architectures*, Conte, T.M. and Gimarc, C.E., Eds., Kluwer Academic Publishers, 1995, chap. 2.
2. Dinero IV cache simulator, online at: http://www.cs.wisc.edu/~markhill/DineroIV.
3. Johnson, E. et al., Lossless trace compression, *IEEE Transactions on Computers*, 50(2), 158, 2001.
4. Luo, Y. and John, L.K., Locality based on-line trace compression, *IEEE Transactions on Computers*, 53, June 2004.
5. Bose, P. and Conte, T.M, Performance analysis and its impact on design, *IEEE Computer*, May, 41, 1998.
6. Crowley, P. and Baer, J.-L., On the use of trace sampling for architectural studies of desktop applications, in *Proc. 1st Workshop on Workload Characterization*. Also in *Workload Characterization: Methodology and Case Studies*, ISBN 0-7695-0450-7, John and Maynard, Eds., IEEE CS Press, 1999, chap. 15.
7. Conte, T.M., Hirsch, M.A., and Menezes, K.N., Reducing state loss for effective trace sampling of superscalar processors, in *Proc. Int. Conf. on Computer Design (ICCD)*, 1996, 468.
8. Skadron, K. et al., Branch prediction, instruction-window size, and cache size: performance tradeoffs and simulation techniques, *IEEE Transactions on Computers*, 48(11), 1260, 1999.
9. Larus, J.R., Efficient program tracing, *IEEE Computer*, May, 52, 1993.
10. Bhargava, R., John, L. K. and Matus, F., Accurately modeling speculative instruction fetching in trace-driven simulation, in *Proc. IEEE Performance, Computers, and Communications Conf. (IPCCC)*, 1999, 65.
11. Moudgill, M., Wellman, J.-D., Moreno, J.H., An approach for quantifying the impact of not simulating mispredicted paths, in *Digest of the Workshop on Performance Analysis and Its Impact on Design (PAID)*, conducted in conjunction with *ISCA 98*.
12. Bechem, C. et al., An integrated functional performance simulator, *IEEE Micro*, 19(3), May 1999.
13. The SimpleScalar Simulator Suite, online at http://www.cs.wisc.edu/~mscalar/simplescalar.html.
14. Boothe, B., Execution driven simulation of shared Memory multiprocessors, in *Fast Simulation of Computer Architectures*, Conte, T.M. and Gimarc, C.E., Eds., Kluwer Academic Publishers, 1995, chap. 6.
15. Emer, J. et al. ASIM: A performance model framework, *IEEE Computer*, 35(2), 68, 2002.
16. Vachharajani, M. et al., Microarchitectural exploration with Liberty, in *Proc. 35th Annual ACM/IEEE Int. Symp. Microarchitecture*, Istanbul, Turkey, November 18–22, 271, 2002.
17. Perez, D., et al., MicroLib: A case for quantitative comparison of microarchitecture mechanisms, in *Proc. MICRO 2004*, Dec 2004.
18. The TRIMARAN home page, online at: http://www.trimaran.org.

19. Manjikian, N., Multiprocessor enhancements of the SimpleScalar tool set, *SIGARCH Computer Architecture News*, 29(1), 8, 2001.
20. Rajwar, R. and Goodman, J., Speculative lock elision: Enabling highly concurrent multithreaded execution, in Proc. *Annual Int. symp. on Microarchtecture* 2001, pp. 294.
21. The SimOS complete system simulator, online at: http://simos.stanford.edu/.
22. The SIMICS simulator, VIRTUTECH. online at: http://www.virtutech.com. Also at: http://www.simics.com/.
23. Shafi, H. et al., Design and validation of a performance and power simulator for PowerPC systems, *IBM Journal of Research and Development*, 47, 5/6, 2003.
24. Reilly, M. and Edmondson, J. Performance simulation of an Alpha microprocessor, *IEEE Computer*, May, 59, 1998.
25. John, L.K. and Liu, Y.-C., A performance model for prioritized multiple-bus multiprocessor systems, *IEEE Transactions on Computers*, 45(5), 580, 1996.
26. Oskin, M., Chong, F.T., and Farrens, M., HLS: Combining statistical and symbolic simulation to guide microprocessor design, in *Proc. Int. Symp. Computer Architecture (ISCA) 27*, 2000, 71.
27. Eeckhout, L. et al., Control flow modeling in statistical simulation for accurate and efficient processor design studies, in *Proc. Int. Symp. Computer Architecture (ISCA)*, 2004.
28. Bell Jr., R.H., et al., Deconstructing and improving statistical simulation in HLS, in *Proc. 3rd Annual Workshop Duplicating, Deconstructing, and Debunking (WDDD)*, 2004.
29. Smith, M., Tracing with Pixie, Report CSL-TR-91-497, Center for Integrated Systems, Stanford University, Nov 1991.
30. Conte, T.M. and Gimarc, C.E., *Fast Simulation of Computer Architectures*, Kluwer Academic Publishers, 1995, chap.3.
31. Srivastava, A. and Eustace, A., ATOM: A system for building customized program analysis tools, in *Proc. SIGPLAN 1994 Conf. on Programming Language Design and Implementation*, Orlando, FL, June 1994, 196.
32. Larus, J., Abstract execution: A technique for efficiently tracing programs, *Software Practice and Experience*, 20(12), 1241, 1990.
33. The PIN program instrumentation tool, online at: http://www.intel.com/cd/ids/developer/asmo-na/eng/183095.htm.
34. The Jaba profiling tool, online at: http://www.ece.utexas.edu/projects/ece/lca/jaba.html.
35. Radhakrishnan, R., Rubio, J., and John, L.K., Characterization of java applications at Bytecode and Ultra-SPARC machine code levels, in *Proc. IEEE Int. Conf. Computer Design*, 281.
36. Heidelberger, P. and Lavenberg, S.S., Computer performance evaluation methodology, in *Proc. IEEE Transactions on Computers*, 1195, 1984.
37. Noonburg, D.B. and Shen, J.P., A framework for statistical modeling of superscalar processor performance, in *Proc. 3rd Int. Symp. High Performance Computer Architecture (HPCA)*, 1997, 298.
38. Sorin, D.J. et al., Analytic evaluation of shared memory systems with ILP processors, in *Proc. Int. Symp. Computer Architecture*, 1998, 380.
39. Karkhanis and Smith, A., first order superscalar processor model, in *Proc. 31st Int. Symp. Computer Architecture*, June 2004, 338.
40. Clark, M. and John, L.K., Performance evaluation of configurable hardware features on the AMD-K5, in *Proc. IEEE Int. Conf. Computer Design*, 1999, 102.

41. Dean, J. et al., Profile me: Hardware support for instruction level profiling on out of order processors, in *Proc. MICRO-30*, 1997, 292.

42. Bhandarkar, D. and Ding, J., Performance characterization of the PentiumPro processor, in *Proc. 3rd High Performance Computer Architecture Symp.*, 288. 1997,

43. Luo, Y. et al. Benchmarking internet servers on superscalar machines. *IEEE Computer*, February, 34, 2003.

44. Vtune, online at: http://www.intel.com/software/products/vtune/.

45. PMON, online at: http://www.ece.utexas.edu/projects/ece/lca/pmon.

46. The Brink Abyss tool for Pentium 4, online at: http://www.eg.bucknell.edu/ ~bsprunt/emon/brink_abyss/brink_abyss.shtm.

47. Merten, M.C. et al., A hardware-driven profiling scheme for identifying hot spots to support runtime optimization, in *Proc. 26th Int. Symp. Computer Architecture*, 1999, 136.

48. Bhargava, R. et al., Understanding the impact of x86/NT computing on microarchitecture, Paper ISBN 0-7923-7315-4, in *Characterization of Contemporary Workloads*, Kluwer Academic Publishers, 2001, 203.

49. Poursepanj, A. and Christie, D., Generation of 3D graphics workload for system performance analysis, in *Proc. 1st Workshop Workload Characterization*. Also in *Workload Characterization: Methodology and Case Studies*, John and Maynard, Eds., IEEE CS Press, 1999.

50. Agarwal, A., Sites, R.L. and Horowitz, M., ATUM: A new technique for capturing address traces using microcode, in *Proc. 13th Int. Symp. Computer Architecture*, 1986, 119.

51. Brooks, D. et al., Wattch: A framework for architectural-level power analysis and optimizations, in *Proc. 27th Int. Symp. Computer Architecture (ISCA)*, Vancouver, British Columbia, June 2000.

52. Gurumurthi, S. et al., Using complete machine simulation for software power estimation: The SoftWatt approach, in *Proc. 2002 Int. Symp. High Performance Computer Architecture*, 2002, 141.

53. The POWER-IMPACT simulator, online at: http://eda.ee.ucla.edu/PowerImpact/main.html.

54. Shivakumar, P. and Jouppi, N.P., CACTI 3.0: An integrated cache timing, power, and area model, Report WRL-2001-2, Digital Western Research Lab (Compaq), Dec 2001.

55. The HotLeakage leakage power simulation tool, online at: http://lava.cs.virginia.edu/HotLeakage/.

56. Black, B. and Shen, J.P., Calibration of microprocessor performance models, *IEEE Computer*, May, 59, 1998.

57. Gibson, J. et al., FLASH vs. (Simulated) FLASH: Closing the simulation loop, in *Proc. 9th Int. Conf. Architectural Support for Programming Languages and Operating Systems*, Cambridge, Massachusetts, United States, Nov 2000, 49.

58. Desikan, R. et al., Measuring Experimental Error in Microprocessor Simulation, in *Proc. 28th Annual Int. Symp. Computer Architecture*, Sweden, June 2001, 266.

59. The WorldWide Computer Architecture home page, Tools Link, online at: http://www.cs.wisc.edu/~arch/www/tools.html.

Chapter Three

Benchmarks

Lizy Kurian John

Contents

Benchmarks used for performance evaluation of computers should be representative of applications that run on actual systems. Contemporary computer workloads include a variety of applications, and it is not easy to define or create representative benchmarks. Performance evaluation using benchmarks has always been a controversial issue for this reason.

It is easy to understand that different benchmarks are appropriate for systems targeted for different purposes. However, it is also a fact that single and simple numbers are easy to understand. One might notice that, even today, many buy their computers based on their clock frequency or memory capacity as opposed to any results based on any benchmark applications.

Three or four decades ago, speed of an ADD instruction or a MULTIPLY instruction was used as an indicator of a computer's performance. Then, there were microbenchmarks and synthetic programs. In the 1980s, computer performance was typically evaluated with small benchmarks, such as kernels, extracted from applications (e.g., Lawrence Livermore Loops, Linpack, Sorting, Sieve of Eratosthenes, 8-queens problem, Tower of Hanoi) or synthetic programs such as Whetstone or Dhrystone[1]. Whetstone was a synthetic floating-point benchmark crafted after studying several floating-point programs. Dhrystone was created with the same philosophy to measure integer performance. Both of these programs were very popular for many years. They both were simple programs and were efforts to create a typical or average program based on the characteristics of many programs. The programs actually did not compute anything useful, and many results computed during the program's run were not ever printed or used. Hence it was easy for optimizing compilers to remove a large part of the code during *dead-code elimination*. Weicker's 1990 paper [1] provides a good characterization of these and other simple benchmarks. Misuse/abuse has happened in the use of these programs and in interpretation of results from these programs. Synthetic benchmarks have been in disrepute since then. The Standard Performance Evaluation Cooperative (SPEC) consortium [2] and the Transactions Processing Council (TPC) [3] formed in 1988 have made available several benchmark suites and benchmarking guidelines to improve the quality of benchmarking.

Benchmarks can be of different types. Many popular benchmarks are programs that perform a fixed amount of computation. The computer that performs the task in the minimum amount of time is considered the winner. There are also throughput benchmarks, in which there is no concept of finishing the fixed amount of work. Throughput benchmarks are used to measure the rate at which work gets done, that is, a task accomplished in a fixed time is used to compare processors or systems. The SPEC CPU benchmarks are examples of fixed-computation benchmarks, whereas the TPC benchmarks are examples of throughput benchmarks. One may also design benchmarks where neither computation nor time is kept fixed. The HINT benchmark [4,5] explained in Section 3.10 is an example of such a benchmark.

This chapter discusses benchmarks from various domains of computers and microprocessors. Table 3.1 lists several popular benchmarks for different classes of workloads. Scientific and technical workloads are computation and memory intensive. They are also called CPU benchmarks. Embedded system applications may have requirements on reducing memory consumption. Java programs are used from embedded systems to servers, and they are shown in a category by themselves. Personal computer applications also form their own category, with their emphasis on word processor, spreadsheet, and other applications for the masses.

3.1 CPU benchmarks

3.1.1 SPEC CPU Benchmarks

SPEC CPU2000 is the current industry standard for CPU-intensive benchmarks. A new suite is expected in 2005. The SPEC [2] was founded in 1988 by a small number of workstation vendors who realized that the marketplace was in desperate need of realistic, standardized performance tests. The basic SPEC methodology is to provide the benchmarker with a standardized suite of source code that comes from existing applications and has already been ported to a wide variety of platforms by its membership. The benchmarker then takes this source code and compiles it for the system in evaluation. The use of already accepted and ported source code greatly reduces the problem of making apples-to-oranges comparisons. SPEC designed CPU2000 to provide a comparative measure of compute-intensive performance across the widest practical range of hardware. The SPEC philosophy has resulted in source code benchmarks developed from real user applications. These benchmarks measure the performance of the processor, memory, and compiler on the tested system.

The SPEC CPU2000 suite contains 14 floating-point programs (4 written in C and 8 in Fortran) and 12 integer programs (11 written in C and 1 in C++). Table 3.2 provides a list of the benchmarks in this suite. The SPEC CPU2000 benchmarks replace the SPEC89, SPEC92, and SPEC95 benchmarks. The SPEC suite contains several input data sets for each program. The reference input set is a large input set, whereas smaller inputs called *test* and *train* are available to test the running environment or to perform profile-based training. Researchers

Table 3.1 Popular Benchmarks for Different Categories of Workloads

Workload Category		Example Benchmark Suite
CPU Benchmarks	Uniprocessor	SPEC CPU 2000 [2]
		PERFECT CLUB [7]
		Java Grande Forum Benchmarks [9]
		SciMark [10]
		ASCI [11]
	Parallel processor	SPLASH [12,13]
		NPB (NAS Parallel Benchmarks) [14]
		ASCI [11]
Multimedia Embedded Systems Digital Signal Processing		EEMBC benchmarks [15]
		BDTI benchmarks [16]
		MediaBench [17]
		MiBench [18]
Java	Client side	SPECjvm98 [2]
		CaffeineMark [21]
		Morphmark [22]
	Server side	SPECjBB2000 [2]
		VolanoMark [23]
	Scientific	Java Grande Forum Benchmarks [9]
		SciMark [10]
Transaction Processing	OLTP (Online transaction processing)	TPC-C [3]
		TPC-W [3]
	DSS (Decision support system)	TPC-H [3]
		TPC-R [3]
Web Server		SPEC web99 [2]
		TPC-W [3]
		VolanoMark [23]
Electronic Commerce	With commercial database	TPC-W [3]
	Without database	SPECjBB2000 [2]
Mail Server		SPECmail2001, IMAP2003 [2]
Network File System		SPEC SFS3.0/LADDIS [2]
Personal Computer		SYSMARK [25]
		Ziff Davis WinBench [24]
		MacBench [27]

at University of Minnesota also created a set called *MinneSPEC* [6] with miniature inputs that are smaller than the reference input sets. Tables 3.3, 3.4, and 3.5 list the programs in the various retired SPEC CPU suites, along with the source languages, and illustrate the evolution of the benchmarks. The SPEC89 suite contained 4 integer programs written in C and 6 floating-point programs written in Fortran. The SPEC92 suite had more programs and included some floating-point programs written in C. The SPEC95 suite

Table 3.2 Programs in the SPEC CPU 2000 Suite

Program	Application	INT/FP	Language	Input	Dynamic Instrn Count
Gzip	Compression	INT	C	input.graphic	103.7 billion
vpr	FPGA placement and routing	INT	C	route	84.06 billion
gcc	C compiler	INT	C	166.i	46.9 billion
mcf	Combinatorial optimization	INT	C	inp.in	61.8 billion
crafty	Game playing: chess	INT	C	crafty.in	191.8 billion
parser	Word processing	INT	C		546.7 billion
eon	Computer visualization	INT	C++	cook	80.6 billion
perlbmk	Perl programming language	INT	C	*	*
vortex	Object-oriented database	INT	C	lendian1.raw	118.9 billion
gap	Group theory, interpreter	INT	C	ref.in	269.0 billion
bzip2	Compression	INT	C	input.graphic	128.7 billion
twolf	Place and route simulator	INT	C	ref	346.4 billion
swim	Shallow water modeling	FP	Fortran	swim.in	225.8 billion
wupwise	Physics/quantum chromodynamics	FP	Fortran	wupwise.in	349.6 billion
mgrid	Multigrid solver: 3-D potential field	FP	Fortran	mgrid.in	419.1 billion
mesa	3-D graphics library	FP	C	mesa.in	141.86 billion
galgel	Computational fluid dynamics	FP	Fortran	galgel.in	409.3 billion
art	Image recognition/neural networks	FP	C	C756hel.in	45.0 billion
equake	Seismic wave propagation simulation	FP	C	inp.in	131.5 billion
ammp	Computational chemistry	FP	C	ammp.in	326.5 billion
lucas	Number theory/primality testing	FP	Fortran	lucas2.in	142.4 billion
fma3d	Finite-element crash simulation	FP	Fortran	fma3d.in	268.3 billion
apsi	Meterology: pollutant distribution	FP	Fortran	apsi.in	347.9 billion
applu	Parabolic/elliptic partial differential equations	FP	Fortran	applu.in	223.8 billion
facerec	Image processing: face recognition	FP	Fortran	*	*
sixtrack	High-energy nuclear physics accelerator design	FP	Fortran	*	*

Table 3.3 Program in the SPEC CPU 89 Suite

Program	Application	INT/FP	Language	Input	Dynamic Instrn Count
espresso	Logic minimization	INT	C	bca.in	0.5 billion
li	Lisp interpreter	INT	C	li-input.lsp	7 billion
eqntott	Boolean equation to truth table converter	INT	C	*	*
gcc	G compiler	INT	C	*	*
spice2g6	Analog circuit simulator	FP	Fortran	*	*
doduc	Nuclear reactor model	FP	Fortran	doducin	1.03 billion
fpppp	Quantum chemistry/electron integrals	FP	Fortran	natoms	1.17 billion
matrix300	Saxpy on matrices	FP	Fortran	—	1.9 billion
nasa7	Seven kernels from NASA	FP	Fortran	—	6.2 billion
tomcatv	Vectorized mesh generation	FP	Fortran	—	1 billion

contained 8 integer programs written in C and 10 floating-point programs written in Fortran. The SPEC2000 suite contains 26 programs, including a C++ program. The length of the SPEC CPU benchmarks have increased tremendously, as demonstrated by the instruction counts in Tables 3.2 through 3.5. These instruction counts are based on Alpha binaries for the SimpleScalar simulator. The specific input set used for the experiment is indicated.

3.1.2   PERFECT CLUB benchmarks

Researchers from University of Illinois created a suite with 13 complete Fortran applications [7] from traditional scientific/engineering domains such as fluid dynamics, signal processing, and modeling. The programs have been used for studying high-performance computing systems and to study compiler transformations that can be applied to regular structured programs. Many computation applications in these domains have moved into languages other than Fortran, and this suite is not very commonly used nowadays.

3.1.3   Java grande forum benchmark suite

The Java Grande Forum Benchmark suite consists of three groups of benchmarks; microbenchmarks that test individual low-level operations (e.g., arithmetic, cast, create); Kernel benchmarks, which are the heart of the algorithms of commonly used applications (e.g., heapsort, encryption/decryption, FFT, Sparse matrix multiplication); and applications (e.g., Raytracer, Monte-Carlo simulation, Euler equation solution, Molecular dynamics, etc) [8,9]. These are computation-intensive benchmarks available in Java.

Table 3.4 Programs in the SPEC CPU 92 Suite

Program	Application	INT/FP	Language	Input	Dynamic Instrn Count
espresso	Logic minimization	INT	C	bca.in	0.5 billion
li	Lisp interpreter	INT	C	li-input.lsp	6.8 billion
eqntott	Boolean equation to truth table converter	INT	C	*	*
compress	Lempel Ziv compression	INT	C	in	0.1 billion
sc	Budgets and spreadsheets	INT	C	*	*
gcc	C compiler	INT	C	*	*
spice2g6	Analog circuit simulator	FP	Fortran	*	*
doduc	Nuclear reactor model	FP	Fortran	doducin	1.03 billion
mdljdp2	Atom motion equation solver (double precision)	FP	Fortran	input.file	2.55 billion
mdljsp2	Atom motion equation solver (single precision)	FP	Fortran	input.file	3.05 billion
wave5	Maxwell equations	FP	Fortran	—	3.53 billion
hydro2d	Navier Stokes equations for hydrodynamics	FP	Fortran	hydro2d.in	44 billion
Swm256	Shallow water equations	FP	Fortran	swm256.in	10.2 billion
alvinn	Neural network	FP	C	In_pats.txt	4.69 billion
ora	Ray tracing	FP	Fortran	params	4.72 billion
ear	Sound to cochleogram by FFTs and math library	FP	C	*	*
su2cor	Particle simulation	FP	Fortran	su2cor.in	4.65 billion
fpppp	Quantum chemistry/ electron integrals	FP	Fortran	natoms	116 billion
nasa7	Seven kernels from NASA	FP	Fortran	—	6.23 billion
tomcatv	Vectorized mesh generation	FP	Fortran	—	0.9 billion

3.1.4 SciMark

SciMark is a composite Java benchmark measuring the performance of numerical codes occurring in scientific and engineering applications [10]. It consists of five computational kernels: FFT, Gauss-Seidel relaxation, sparse matrix-multiply, Monte-Carlo integration, and dense LU factorization. These kernels are selected to provide an indication of how well the underlying Java Virtual Machines (JVMs) perform on applications that utilize these types of algorithms. The problem sizes are purposely made small in order to isolate the effects of memory hierarchy and focus on internal JVM/JIT and CPU issues. A larger version of the benchmark (SciMark 2.0 LARGE) addresses performance of the memory subsystem with out-of-cache problem sizes.

Table 3.5 Programs in the SPEC CPU 95 Suite

Program	Application	INT/FP	Language	Input	Dynamic Instrn Count
go	Go-playing	INT	C	null.in	18.2 billlion
li	Xlisp interpreter	INT	C	*.lsp	75.6 billion
m88ksim	Chip simulator	INT	C	ctl.in	520.4 billion
compress	Unix compress	INT	C	bigtest.in	69.3 billion
ijpeg	Image compression/ decompression	INT	C	penguin.ppm	41.4 billion
gcc	GNU C compiler	INT	C	expr.i	1.1 billion
perl	Interpreter for Perl	INT	C	perl.in	16.8 billion
vortex	Object-oriented database	INT	C	*	*
wave5		FP	Fortran	wave5.in	30 billion
hydro2d	Navier Stokes equations	FP	Fortran	hydro2d.in	44 billion
swim	Shallow water equations	FP	Fortran	swim.in	30.1 billion
applu	Partial differential equations	FP	Fortran	applu.in	43.7 billion
mgrid	3-D potential field	FP	Fortran	mgrid.in	56.4 billion
Turb3d	Turbulence modeling	FP	Fortran	turb3d.in	91.9
Su2cor	Monte-Carlo method	FP	Fortran	su2cor.in	33 billion
fpppp	Quantum chemistry	FP	Fortran	natmos.in	116 billion
apsi	Weather prediction	FP	Fortran	apsi.in	28.9 billion
tomcatv	Vectorized mesh generator	FP	Fortran	tomcatv.in	26.3 billion

3.1.5   ASCI

The Accelerated Strategic Computing Initiative (ASCI) from Lawrence Livermore Laboratories released several numeric codes suitable for evaluation of compute intensive systems. The programs are available from the ASCI benchmarks Web site [11]. Many of these programs are written to exploit explicit threads for testing multiprocessor systems and message-passing mechanisms. The programs include SPPM (Simplified Piecewise Parabolic Method, which solves a 3-D gas dynamics problem), SWEEP3D (a 3-D Discrete Ordinates Neutron Transport problem), and COMOPS (an inter-SMP communications benchmark that tests nearest-neighbor point-to-point send/receive, also known as ping-pong, as well as 2-D ghost cell update, 3-D ghost cell update, broadcast, reduce, gather, and scatter operations). Parallel versions of the benchmarks can be created for parallel processor benchmarking. The 24 Lawrence Livermore kernels are also available from ASCI [11]. Although they have been in disrepute in recent years, they contain examples of computations in various popular scientific applications. Some of these loops are very parallel, whereas some illustrate the typical type of intra-loop and inter-loop dependencies.

3.1.6 SPLASH

Stanford researchers [12,13] created the SPLASH suite for parallel-processor benchmarking. The first SPLASH suite consisted of six scientific and engineering applications, and the SPLASH2 suite contains eight complete applications and four kernels. These programs represent a variety of computations in scientific, engineering, and graphics computing. The application programs in the SPLASH2 suite are Barnes (galaxy/particle simulation), FMM (particle simulation similar to Barnes in functionality but contains more unstructured communication patterns), Ocean (ocean current simulation), Radiosity (iterative hierarchical diffuse radiosity method), Raytrace (3-D scene ray tracing), Volrend (volume rendering using ray casting), Water-Nsquared (forces in water), and Water spatial (forces in water problem using 3-D grid of cells algorithm). SPLASH2 also contain four kernels: Radix (radix sort kernel), Cholesky (blocked sparse Cholesky factorization kernel), FFT (six-step FFT kernel optimized to minimize communication), and LU (matrix factorization kernel).

3.1.7 NAS parallel benchmarks

The NAS Parallel Benchmarks (NPB) [14] are a set of programs designed to help evaluate the performance of parallel supercomputers. The early version of the benchmarks consisted of kernels/pseudo applications derived from computational fluid dynamics (CFD) applications. NPB includes programs with different types of parallelism and dependency characteristics. Recent releases include NPB3, which contains parallel implementations of the programs using OpenMP, High Performance Fortran (HPF), and Java, respectively. They were derived from the previous NPB-serial implementations after some additional optimization. Another recent release is GridNPB 3, which is a new suite of benchmarks designed specifically to rate the performance of computational grids. Each of the four benchmarks consists of a collection of communicating tasks derived from the NPB. They symbolize distributed applications typically run on grids. The distribution contains serial and concurrent reference implementations in Fortran and Java.

3.2 Embedded and media benchmarks

3.2.1 EEMBC benchmarks

The EDN Embedded Microprocessor Benchmark Consortium (EEMBC, pronounced "embassy") was formed in April 1997 to develop meaningful performance benchmarks for processors in embedded applications. The EEMBC benchmarks [15] comprise a suite of benchmarks designed to reflect real-world applications as well as some synthetic benchmarks. These benchmarks target the automotive/industrial, consumer, networking, office automation, and telecommunications markets. More specifically, these benchmarks target specific applications that include engine control, digital cameras, printers, cellular phones, modems, and similar devices with

Table 3.6 Programs in the EEMBC Suite

Category	Program
1. Automotive/Industrial	Angle-to-Time Conversion
	Basic integer and floating point
	Bit manipulation
	Cache buster
	CAN remote data request
	Fast Fourier transform (FFT)
	Finite impulse response (FIR) filter
	Inverse discrete cosine transform (IDCT)
	Inverse fast Fourier transform (iFFT)
	Infinite impulse response (IIR) filter
	Matrix arithmetic
	Pointer chasing
	Pulse width modulation (PWM)
	Road speed calculation
	Table lookup and interpolation
	Tooth to spark
2. Consumer	High-pass gray-scale filter
	JPEG
	RGB to CMYK conversion
	RGB to YIQ conversion
3. GrinderBench for the Java 2 Micro Edition Platform	Chess
	Cryptography
	kXML
	ParallelBench
	PNG decoding
	Regular expression
4. Networking	Packet flow
	OSPF
	Router lookup
5. Office Automation	Dithering
	Image rotation
	Text processing
6. Telecom	Autocorrelation
	Bit allocation
	Convolutional encoder
	Fast Fourier transform (FFT)
	Viterbi decoder

embedded microprocessors. The EEMBC consortium dissected applications from these domains and derived 37 individual algorithms that constitute the EEMBC's Version 1.0 suite of benchmarks. The programs in the suite are listed in Table 3.6. EEMBC establishes benchmark standards and provides certified benchmarking results through the EEMBC Certification Labs (ECL) in Texas and California. EEMBC is backed by the majority of the processor industry and has therefore established itself as the industry-standard embedded processor benchmarking forum.

3.2.2 BDTI benchmarks

Berkeley Design Technology, Inc. (BDTI) is a technical services company that has focused exclusively on digital signal processing since 1991. BDTI provides the industry standard BDTI Benchmarks™, a proprietary suite of DSP benchmarks [16]. BDTI also develops custom benchmarks to determine performance on specific applications. The benchmarks contain DSP routines such as FIR filter, IIR filter, FFT, dot product, and Viterbi decoder.

3.2.3 MediaBench

The MediaBench benchmark suite consists of several applications belonging to the image processing, communications, and DSP applications. Examples of applications that are included are JPEG, MPEG, GSM, G.721 Voice compression, Ghostscript, and ADPCM. JPEG is the compression program for images, MPEG involves encoding/decoding for video transmission, Ghostscript is an interpreter for the Postscript language, and ADPCM is adaptive differential pulse code modulation. The MediaBench is an academic effort to assemble several media processing related benchmarks. An example of the use of these benchmarks may be found in the proceedings of the thirtieth International Symposium on Microarchitecture [17].

3.2.4 MiBench

MiBench is a free embedded benchmark suite [18,19], with similar programs as the EEMBC suite. The EEMBC suite is not readily accessible to academic researchers. To solve this problem, researchers at Michigan compiled a set of 35 embedded programs to form the MiBench suite. Modeled around the EEMBC suite, programs are grouped into six categories: automotive, consumer, network, office, security, and telecommunications. All the programs are available as C source code. Embedded benchmarks were written in assembly a few years ago; however, the current trend in the embedded domain has been to use compilers and C source code. MiBench can be ported to any embedded platform because its source code is available.

3.3 Java benchmarks

3.3.1 SPECjvm98

The SPECjvm98 suite consists of a set of programs intended to evaluate performance for the combined hardware (CPU, cache, memory, and other platform-specific performance) and software aspects (efficiency of JVM, the JIT compiler, and OS implementations) of the JVM client platform [2]. The SPECjvm98 uses common computing features, such as integer and floating-point operations, library calls, and I/O, but does not include AWT (window), networking, and graphics. Each benchmark can be executed with three different input sizes referred to as S1, S10, and S100. The seven programs are compression/decompression (compress), expert system (jess), database (db), Java compiler (javac), mpeg3 decoder (mpegaudio), raytracer (mtrt), and a parser (jack).

3.3.2   SPECjbb2000

Java Business Benchmark (JBB) is SPEC's first benchmark for evaluating the performance of server-side Java. The benchmark emulates an electronic commerce workload in a three-tier system. It is written is Java, adapting a portable business-oriented benchmark called pBOB, written by IBM. Although it is a benchmark that emulates business transactions, it is very different from the TPC benchmarks. There are no actual clients; they are replaced by driver threads. Similarly, there is no actual database access. Data is stored as binary trees of objects. The benchmark contains business logic and object manipulation, primarily representing the activities of the middle tier in an actual business server. The SPECjbb allows configuration of the number of warehouses to create scalable benchmarks. The number of warehouses was fixed at 10 and 25 in a study on IBM and Intel processors [20].

3.3.4   CaffeineMark

The CaffeineMark was a series of benchmarks to help gauge performance of Java on the Internet. The benchmark suite analyzes Java system performance in 11 different areas, 9 of which can be executed directly over the Internet. It is almost the industry-standard Java benchmark. The Caffeine-Mark was widely used for comparing different JVMs on a single system; that is, it compared applet viewers, interpreters and JIT compilers from different vendors. The CaffeineMark benchmark was used as a measure of Java applet/application performance across platforms. CaffeineMark 2.5 and CaffeineMark 3.0 [21] have been used in Java system benchmarking.

3.3.5   MorphMark

Games are becoming an increasingly important workload on mobile phones. MorphMark [22] performs a series of tests to determine which Java-enabled mobile handsets are best suited to run games. The MorphMark suite tests the performance of the JVM, the graphics on the handset, Java I/O performance, and similar performance aspects.

3.3.6   VolanoMark

VolanoMark is a pure Java server benchmark with long-lasting network connections and high thread counts [23]. It can be divided into two parts — server and client — although they are provided in one package. It is based on a commercial chat server application, the VolanoChat, which is used in several countries worldwide. The server accepts connections from the chat client. The chat client simulates many chat rooms and many users in each chat room. The client continuously sends messages to the server and waits for the server to broadcast the messages to the users in the same chat room. VolanoMark creates two threads for each client connection. VolanoMark can be used to test both speed and scalability of a system. In the speed test, it is executed in an iterative fashion on a single machine. In scalability test, the server and client are executed on separate machines with high-speed network connection.

3.3.7 SciMark

See Section 3.1.3 earlier in this chapter.

3.3.8 Java grande forum benchmarks

See Section 3.1.2 earlier in this chapter.

3.4 Transaction processing benchmarks

The Transaction Processing Council (TPC) [3] is a nonprofit corporation that was founded in 1988 to define transaction processing and database benchmarks and to disseminate objective, verifiable transaction-processing performance data to the industry. The term transaction is often applied to a wide variety of business and computer functions. When viewed as a computer function, a transaction could refer to a set of operations including disk accesses, operating system calls, or some form of data transfer from one subsystem to another. TPC regards a transaction as it is commonly understood in the business world: a commercial exchange of goods, services, or money. A typical transaction, as defined by the TPC, would include the updating to a database system for such things as inventory control (goods), airline reservations (services), or banking (money). In these environments, a number of customers or service representatives input and manage their transactions via a terminal or desktop computer connected to a database. Typically, the TPC produces benchmarks that measure transaction processing (TP) and database (DB) performance in terms of how many transactions a given system and database can perform per unit of time, for example, transactions per second or transactions per minute. The TPC benchmarks can be classified into two categories: Online Transaction Processing (OLTP) and Decision Support Systems (DSSs). OLTP systems are used in day-to-day business operations (airline reservations, banks) and are characterized by large numbers of clients who continually access and update small portions of the database through short-running transactions. DSSs are primarily used for business analysis purposes, to understand business trends, and for guiding future business directions. Information from the OLTP side of the business is periodically fed into the DSS database and analyzed. DSS workloads are characterized by long-running queries that are primarily read-only and may span a large fraction of the database. There are four benchmarks that are active: TPC-C, TPC-W, TPC-R, and TPC-H. These benchmarks can be run with different data sizes, or scale factors. In the smallest case (or scale factor = 1), the data size is approximately 1 gigabyte (GB). The early TPC benchmarks, namely TPC-A, TPC-B, and TPC-D, have become obsolete.

3.4.1   TPC-C

TPC-C is an OLTP benchmark. It simulates a complete computing environment where a population of users executes transactions against a database. The benchmark is centered around the principal activities (transactions) of a business similar to that of a worldwide wholesale supplier. The transactions include entering and delivering orders, recording payments, checking the status of orders, and monitoring the level of stock at the warehouses. Although the benchmark portrays the activity of a wholesale supplier, TPC-C is not limited to the activity of any particular business segment but rather represents any industry that must manage, sell, or distribute a product or service. TPC-C involves a mix of five concurrent transactions of different types and complexity either executed online or queued for deferred execution. There are multiple online terminal sessions. The benchmark can be configured to use any commercial database system, such as Oracle, DB2 (IBM), or Informix. Significant disk input and output are involved. The databases consist of many tables with a wide variety of sizes, attributes, and relationships. The queries result in contention on data accesses and updates. TPC-C performance is measured in new-order transactions per minute (tpmC). The primary metrics are the transaction rate (tpmC) and price per performance metric ($/tpmC).

3.4.2   TPC-H

The TPC Benchmark™H (TPC-H) is a DSS benchmark. As discussed earlier, DSSs are used primarily for analyzing business trends. For instance, a DSS database may consist of records of transactions from the previous several months of a company's operation, which can be analyzed to shape the future strategy of the company. DSS workloads typically involve long-running queries spanning large databases. The TPC-H benchmark consists of a suite of business-oriented ad hoc queries and concurrent data modifications. The queries and the data populating the database have been chosen to have broad industry-wide relevance. This benchmark is modeled after DSSs that examine large volumes of data, execute queries with a high degree of complexity, and give answers to critical business questions. There are 22 queries in the benchmark. These involve database operations such as scan, indexed scan, select, join, and merge operations. Each of the 22 queries may include multiple database operations. The benchmark involves database tables ranging in size from a few kilobytes to several gigabytes. The benchmark dataset can be scaled to various sizes, and the smallest acceptable size for an auditable

TPC-H system is 1 GB. The performance metric reported by TPC-H is called the TPC-H composite query-per-hour performance metric (QphH@Size), and the TPC-H price-per-performance metric is $/QphH@Size. The queries in TPC-H are the same as the queries in the TPC-R. One may not perform optimizations based on a priori knowledge of queries in TPC-H, whereas TPC-R permits such optimizations.

3.4.3   TPC-R

The TPC Benchmark™R (TPC-R) is a DSS benchmark similar to TPC-H, but it allows additional optimizations based on advance knowledge of the queries. It consists of a suite of business-oriented queries and concurrent data modifications. As in TPC-H, there are 22 queries. The queries and the data tables are the same as those in TPC-H. The performance metric reported by TPC-R is called the TPC-R composite query-per-hour performance metric (QphR@Size), and the TPC-R price-per-performance metric is $/QphR@Size.

3.4.4   TPC-W

TPC Benchmark™ W (TPC-W) is a transactional Web benchmark. The workload simulates the activities of a business-oriented transactional Web server in an electronic commerce environment. It supports many of the features of the TPC-C benchmark and has several additional features related to dynamic page generation with database access and updates. Multiple online browser sessions and online transaction processes are supported. Contention on data accesses and updates are modeled. The performance metric reported by TPC-W is the number of Web interactions processed per second (WIPS). Multiple Web interactions are used to simulate the activity of a retail store, and each interaction is subject to a response time constraint. Different profiles can be simulated by varying the ratio of browsing and buying, that is, it simulates customers who are primarily browsing and those who are primarily shopping.

3.5   Web server benchmarks

3.5.1   SPECweb99

SPECweb99 is the SPEC benchmark for evaluating the performance of World Wide Web Servers [2]. It measures a system's ability to act as a Web server. The initial effort from SPEC in this direction was SPECweb96, but it contained only static workloads, meaning that the requests were for simply downloading Web pages that did not involve any computation. However, if one examines the use of the Web, it is clear that many downloads involve computation to generate the information the client is requesting. Such Web pages are referred to as dynamic Web pages. SPECweb99 includes dynamic Web pages. The file accesses are made to closely match today's real-world

Web server access patterns. The pages also contain dynamic ad rotation using cookies and table lookups.

3.5.2   *VolanoMark*

See Section 3.3.6 earlier in this chapter.

3.5.3   *TPC-W*

See Section 3.4.4 earlier in this chapter.

3.6   *E-Commerce benchmarks*

Electronic commerce (e-commerce) has become very popular in the recent years. A significant amount of merchandise is sold over electronic outlets such as amazon.com. The TPC-W benchmark described in Section 3.4.4 models such an environment. This is one side of e-commerce as it affects the buyers. Managing the business and deciding business strategies is another part of e-business (electronic business). The TPC-H and TPC-R benchmarks model typical activity as performed by corporations in order to successfully conduct their business. The TPC benchmarks require a commercial database program and are difficult to handle in many simulation environments. The SPECjbb2000 benchmark described in Section 3.3.2 is an e-commerce benchmark in which the database has been simplified to data structures in the program as opposed to an actual database.

3.7   *Mail server benchmarks*

3.7.1   *SPECmail2001*

SPECmail2001 is a standardized SPEC mail server benchmark designed to measure a system's ability to act as a mail server servicing e-mail requests. The benchmark characterizes throughput and response time of a mail server system under test with realistic network connections, disk storage, and client workloads. The benchmark focuses on the Internet service provider (ISP) as opposed to Enterprise class of mail servers, with an overall user count in the range of approximately 10,000 to 1,000,000 users. The goal is to enable objective comparisons of mail server products.

3.8   *File server benchmarks*

3.8.1   *System file server version 3.0*

System File Server Version 3.0 (SFS 3.0) is SPEC's benchmark for measuring NFS (Network File System) file server performance across different vendor platforms. It contains a workload that was developed based on a survey of more than 1,000 file servers in different application environments.

3.9 PC benchmarks

Applications on the personal computer (PC) are very different from applications on servers. PC users typically perform such activities as word processing, audio and video applications, graphics, and desktop accounting. A variety of benchmarks are available, primarily from Ziff Davis and BAPCO, to benchmark the Windows-based PC. Ziff Davis [24] Winstone and Bapco SYSMARK [25] are benchmarks that measure overall performance, whereas the other benchmarks are intended to measure performance of one subsystem such as video or audio or one aspect such as power. MacBench [27] is a subsystem-level benchmark that measures the performance of a Macintosh operating system's graphics, disk, processor, FPU, video, and CD-ROM subsystems. Table 3.7 lists the most common PC benchmarks

3.10 The HINT benchmark

The HINT benchmark has a very different philosophy than all the benchmarks described so far. It is a variable-computation, variable-time benchmark. It solves a mathematical integration problem, whose result continually improves as more computations are performed. The system under test continues to work on the problem, and the quality of the solution obtained speaks of the capability of the system.

The benchmark tries to find the upper and lower bounds for

$$(1 - x)/(1 + x)dx$$

A technique called interval subdivision is used to find the answer. The range is divided into a number of intervals, and the answer is computed by counting the number of squares that are contributing to the lower and upper bounds. A better answer can be obtained by splitting the intervals into smaller subintervals. A computer is rated by analyzing the goodness of the answer. Essentially, a computer with more computing and memory capability will be able to generate a better answer to the problem. A metric, QUIPS, based on the quality of the answer has been defined to compare different systems. Whereas fixed-work benchmarks get outdated when computing capability or cache/memory capacity increases, the HINT benchmark automatically scales for larger systems. A more detailed description of the benchmark can be obtained from some of the sources listed at the end of this chapter [4,5,28].

3.11 Return of synthetic benchmarks

Many of the modern benchmarks are very long. It takes prohibitively long periods of time to perform simulations on them. Recent research shows that short synthetic streams of instructions can be created to approximately

Table 3.7 Popular Personal Computer Benchmarks

Benchmark	Description
Business Winstone [24]	A system-level, application-based benchmark that measures a PC's overall performance when running today's top-selling Windows-based 32-bit applications. It runs real business applications through a series of scripted activities and uses the time a PC takes to complete those activities to produce its performance scores. The suite includes five Microsoft Office 2000 applications (Access, Excel, FrontPage, PowerPoint, and Word), Microsoft Project, Lotus Notes R5, NicoMak WinZip, Norton AntiVirus, and Netscape Communicator.
WinBench [24]	A subsystem-level benchmark that measures the performance of a PC's graphics, disk, and video subsystems in a Windows environment.
3DwinBench [24]	Tests the bus used to carry information between the graphics adapter and the processor subsystem. Hardware graphics adapters, drivers, and enhancing technologies such as MMX/SSE are tested.
CD WinBench [24]	Measures the performance of a PC's CD-ROM subsystem, which includes the CD drive, controller, and driver, and the system processor.
Audio WinBench [24]	Measures the performance of a PC's audio subsystem, which includes the sound card and its driver, the processor, the DirectSound and DirectSound 3D software, and the speakers.
Battery Mark [24]	Measures battery life on notebook computers.
I-bench [24]	A comprehensive, cross-platform benchmark that tests the performance and capability of Web clients. The benchmark provides a series of tests that measure both how well the client handles features and the degree to which network access speed affects performance.
Web Bench [24]	Measures Web server software performance by running different Web server packages on the same server hardware or by running a given Web server package on different hardware platforms.
NetBench [24]	A portable benchmark program that measures how well a file server handles file I/O requests from clients. NetBench reports throughput and client response time measurements.
3Dmark [26]	From Futuremark Corporation. A nice 3-D Benchmark, which measures 3-D gaming performance. Results are dependent on CPU, memory architecture, and the 3-D Accelerator employed.

Table 3.7 Popular Personal Computer Benchmarks (Continued)

Benchmark	Description
SYSMARK [25]	Measures a system's real-world performance when running typical business applications. This benchmark suite comprises the retail versions of eight application programs and measures the speed with which the system under test executes predetermined scripts of user tasks typically performed when using these applications. The performance times of the individual applications are weighted and combined into both category-based performance scores as well as a single overall score. The application programs employed by SYSmark 32 are Microsoft Word and Lotus WordPro (for word processing), Microsoft Excel (for spreadsheet), Borland Paradox (for database), CorelDraw (for desktop graphics), Lotus Freelance Graphics and Microsoft PowerPoint (for desktop presentation), and Adobe Pagemaker (for desktop publishing).

match the behavior of the instruction stream from the full execution [29,30]. Synthetic streams as small as 0.1% of the size of the full benchmark are able to capture the essential behavior of the actual execution. Although early synthetic benchmarks such as Whetstone and Dhrystone have been in disrepute, difficulties with long benchmarks may make these synthetic instruction streams useful.

3.12   Conclusion

Benchmark suites are updated very frequently. Those interested in experimental performance evaluation should continuously monitor emerging benchmarks. The Web resources listed at the end of this chapter can provide new information on benchmarks as they become available. Microprocessor vendors are inclined to show off their products in the best light, to project results for benchmarks that run well on their system, and to develop special optimizations within their compilers to obtain improved benchmark scores, while staying within the legal limits of the benchmark guidelines. It is extremely important to understand benchmarks, their features, and the metrics used for performance evaluation in order to correctly interpret the performance results.

References

1. Weicker, Reinhold P., An overview of common benchmarks *IEEE Computer* December 1990, 65–75.
2. SPEC Benchmarks, online at: http://www.spec.org.

3. Transactions Processing Council, online at: http://www.tpc.org.
4. Gustafson, J.L. and Snell, Q.O., HINT: A new way to measure computer performance, Hawaii International Conference on System Sciences, 1995, pp. II: 392–401.
5. HINT, online at: http://www.scl.ameslab.gov/scl/HINT/HINT.html.
6. KleinOsowski, A.J and Lilja, D.J., MinneSPEC: A new SPEC benchmark workload for simulation-based computer architecture research, *Computer Architecture Letters*, Vol. 1, June 2002.
7. The PERFECT CLUB benchmarks, online at: http://www.csrd.uiuc.edu/benchmark/benchmark.html.
8. Mathew, J.A., Coddington, P.D., and Hawick, K.A., Analysis and development of the Java Grande Benchmarks, Proceedings of the ACM 1999 Java Grande Conference, June 1999.
9. Java Grande Benchmarks, online at: http://www.epcc.ed.ac.uk/javagrande/.
10. SciMark, online at: http://math.nist.gov/scimark2.
11. ASCI Benchmarks, online at: http://www.llnl.gov/asci_benchmarks/asci/asci_code_list.html.
12. Singh, J.P., Weber, W.-D., and Gupta, A., SPLASH: Stanford parallel applications for shared memory, *Computer Architecture News*, 20(1), 5–44, 1992.
13. Woo, S.C., Ohara, M., Torrie, E., Singh, J.P, and Gupta, A., The SPLASH-2 programs: Characterization and methodological considerations, Proceedings of the 22nd International Symposium on Computer Architecture, pp. 24–36, June 1995.
14. NAS Parallel Benchmarks, online at: http://www.nas.nasa.gov/Software/NPB/.
15. EEMBC, online at: http://www.eembc.org.
16. BDTI, online at: http://www.bdti.com/.
17. Lee, C., Potkonjak, M., and Smith, W.H.M., MediaBench: A tool for evaluating and synthesizing multimedia and communication systems, Proceedings of the 30th International Symposium on Microarchitecture, pp. 330–335.
18. Guthaus, Matthew R., Ringenberg, Jeffrey S., Ernst, Dan, Austin, Todd M., Mudge, Trevor, and Brown, Richard B., MiBench: A free, commercially representative embedded benchmark suite, *IEEE 4th Annual Workshop on Workload Characterization*, Austin, TX, December 2001.
19. MiBench benchmarks, online at: http://www.eecs.umich.edu/mibench/.
20. Luo, Y., Rubio, J., John, L., Seshadri, P., and Mericas A., Benchmarking Internet Servers on SuperScalar Machines, *IEEE Computer*, February 2003, 34–40.
21. The Caffeine Benchmarks, online at: http://www.benchmarkhq.ru/cm30/.
22. Morphmark Midlet, online at: http://www.morpheme.co.uk/.
23. VolanoMark, online at: http://www.volano.com/benchmarks.html.
24. Ziff Davis Benchmarks, online at: http://www.zdnet.com/etestinglabs/filters/benchmarks.
25. SYSMARK, online at: http://www.bapco.com/.
26. 3D Mark Benchmarks, online at: www.futuremark.com.
27. Macbench, online at: http://www.macspeedzone.com.
28. Lilja, D.J., Measuring computer performance: A practitioner's guide, Cambridge University Press, 2001.

29. Eeckhout, L., Bell, R.H., Jr., Stougie, B., De Bosschere, K., and John, L.K., Control flow modeling in statistical simulation for accurate and efficient processor design studies, Proceedings of the International Symposium on Computer Architecture (ISCA), June 2004.

30. Bell, R.H., Jr. and John, L.K., Experiments in automatic benchmark synthesis technical report TR-040817-01, Laboratory for Computer Architecture, University of Texas at Austin, August 2004.

Chapter Four

Aggregating Performance Metrics Over a Benchmark Suite

Lizy Kurian John

Contents

The topic of finding a single number to summarize overall performance of a computer system over a benchmark suite is continuing to be a difficult issue less than 2 decades after Smith's 1988 paper [1]. Although significant insight into the problem has been provided by Smith [1], Hennessey and Patterson [2], and Cragon [3], the research community still seems to be unclear on the correct mean to use for different performance metrics. How should metrics obtained from individual benchmarks be aggregated to present a summary of the performance over the entire suite? What central tendency measures are valid over the whole benchmark suite for speedup, CPI, IPC, MIPS, MFLOPS, cache miss rates, cache hit rates, branch misprediction rates, and other measurements?

Arithmetic mean has been touted to be appropriate for time-based metrics, whereas harmonic mean is touted to be appropriate for rate-based metrics. Is cache miss rate a rate-based metric and, hence, is harmonic mean appropriate? Geometric mean is a valid measure of central tendency for ratios or dimensionless quantities [3]; however, it is also advised that geometric mean should not be used for summarizing any performance measure [1,4]. Speedup, which is a popular metric in most architecture papers to indicate performance enhancement by the proposed architecture is dimensionless and is a ratio-based measure. What will be an appropriate measure to summarize speedups from individual benchmarks?

It is known that weighted means should be used if the benchmarks are not of equal weight. What does equally weighted mean? Does equal weight mean that each benchmark is run once, that each benchmark is equally likely to be in a workload of the user, that all benchmarks have an equal number of instructions, or that all benchmarks run for equal numbers of cycles? Whenever two machines are compared, there is always the question of whether the benchmarks are equally weighted in the baseline machine or the enhanced machine. And note that both cannot be true unless each benchmark is enhanced equally.

This chapter provides some answers to such questions, in the context of aggregating metrics from individual benchmarks in a benchmark suite. It shows that weighted arithmetic or harmonic mean can be used interchangeably and correctly, if the appropriate weights are applied. Mathematical proofs are provided in the chapter to establish this.

4.1 MIPS as an example

Let's start with MIPS as an example metric. Let's assume that the benchmark suite is composed of n benchmarks, and their individual MIPS are known.

We know that the overall MIPS of the entire suite is the total instruction count in millions divided by the total time taken for execution of the whole benchmark suite. Hence,

$$\text{Overall MIPS} = \frac{\sum_{i=1}^{n} I_i}{\sum_{i=1}^{n} t_i} \tag{4.1}$$

where I_i is the instruction count of each component benchmark (in millions) and t_i is the execution time of each benchmark.

Assume MIPS_i is the MIPS rating of each individual benchmark. The overall MIPS is essentially the MIPS when the n benchmarks are considered as parts of a big application. We find that the overall MIPS of the suite can be obtained by computing a *weighted harmonic mean* (WHM) of the MIPS of the individual benchmarks weighted according to the instruction counts or by computing a *weighted arithmetic mean* (WAM) of the individual MIPS with weights corresponding to the execution times spent in each benchmark in the suite. Let us establish this mathematically.

The weights ω_i of the individual benchmarks according to instruction counts are

$$\frac{I_1}{\sum I_i}, \frac{I_2}{\sum I_i}, \quad \text{and so on}$$

All summations in this chapter are for the n benchmarks as in Equation 4.1, and, hence, for compactness we are going to just use the summation sign from now on. The weights of the individual benchmarks according to execution times (t_i) are

$$\frac{t_1}{\sum t_i}, \frac{t_2}{\sum t_i}, \quad \text{and so on.}$$

Now,

$$\text{WHM with weights corresponding to instruction count} = \frac{1}{\sum \dfrac{\omega_i}{MIPS_i}},$$

where ω_i is the weight of benchmark i according to instruction count

$$= \frac{1}{\dfrac{I_1}{\sum I_i} \cdot \dfrac{1}{MIPS_1} + \dfrac{I_2}{\sum I_i} \cdot \dfrac{1}{MIPS_2} + \cdots}$$

$$= \frac{1}{\dfrac{1}{\sum I_i} \sum \dfrac{I_i}{MIPS_i}}$$

$$= \frac{\sum I_i}{\sum \dfrac{I_i}{MIPS_i}} \tag{4.2}$$

$$= \frac{\sum I_i}{\sum \dfrac{I_i t_i}{I_i}}$$

$$= \frac{\sum I_i}{\sum t_i},$$

which we know is overall MIPS according to Equation 4.1.

Now, it can be seen that the same result can be obtained by taking a weighted arithmetic mean of the individual MIPS with weights corresponding to the execution times spent in each benchmark in the suite.

WAM weighted with $time = \sum \omega t_i \cdot MIPS_i$, where (t_i) is the weights according to execution time

$$= \frac{t_1}{\sum t_i} \cdot MIPS_1 + \frac{t_2}{\sum t_i} \cdot MIPS_2 + \cdots$$

$$= \frac{1}{\sum t_i} \left[t_1 \cdot \frac{I_1}{t_1} + t_2 \cdot \frac{I_2}{t_2} + \cdots \right]$$

$$= \frac{1}{\sum t_i} \left[\sum I_i \right]$$

$$= \frac{\sum I_i}{\sum t_i}$$

$$= \text{Overall MIPS}$$

Thus, if the individual MIPS and the relative weights of instruction counts or execution times are known, the overall MIPS can be computed. Table 4.1 illustrates an example benchmark suite with five benchmarks, their individual instruction counts, individual execution times, and the individual MIPS. Let us calculate the overall MIPS of the suite directly from the overall instruction count and the overall execution time. Because the overall instruction count equals 2000 million, and the overall execution time equals 10 seconds, overall MIPS equals 2000/10, that is 200.

We can also calculate the overall MIPS from the individual MIPS and the weights of the individual benchmarks.

Table 4.1   An Example Benchmark Suite with Five Benchmarks, Their Individual Instruction Counts, Individual Execution Times, and Individual MIPS

Benchmarks	Instruction Count (in million)	Time (sec)	Individual MIPS
1	500	2	250
2	50	1	50
3	200	1	200
4	1000	5	200
5	250	1	250

Weights of the benchmarks with respect to instructions counts are

$$500/2000, 50/2000, 200/2000, 1000/2000, 250/2000$$
that is, 0.25, 0.025, 0.1, 0.5, 0.125

Weights of the benchmarks with respect to time are

$$0.2, 0.1, 0.1, 0.5, 0.1$$

WHM of individual MIPS (weighted with I-counts)

$$= 1/(0.25/250 + 0.025/50 + 0.1/200 + 0.5/200 + 0.125/250)$$
$$= 200$$

WAM of individual MIPS (weighted with time)

$$= 250*0.2 + 50*0.1 + 200*0.1 + 200*0.5 + 250*0.1$$
$$= 200$$

Thus, either WAM or WHM can be used to find overall means, if the appropriate weights can be properly applied. It can also be seen that the simple (unweighted) arithmetic mean or simple (unweighted) harmonic mean are not correct, if the target workload is the sum of the five component benchmarks.

Unweighted arithmetic mean of individual MIPS = 190
Unweighted harmonic mean of individual MIPS = 131.58

Neither of these numbers is indicative of the overall MIPS. Of course, the benchmarks are not equally weighted in the suite (either by instruction count or execution time), and hence the unweighted means are not correct.

In general, if a metric is obtained by dividing A by B, and if A is weighed equally among the benchmarks in a suite, harmonic mean is correct. If B is weighed equally among the component benchmarks in a suite, arithmetic mean is correct while calculating the central tendency of the metric obtained by (A/B). In other words, either harmonic mean with weights corresponding to the measure in the numerator or arithmetic mean with weights corresponding to the measure in the denominator is valid when trying to find the aggregate measure from the values of the measures in the individual benchmarks. We use this principle to find the correct means for a variety of performance metrics. This is shown in Table 4.2.

Somehow there seems to be an impression that arithmetic mean is naïve and useless. Arithmetic mean is meaningless for MIPS or MFLOPS when each benchmark contains equal number of instructions or equal number of floating-point operations; however, it is meaningful in many other situations. Consider the following situation: A computer runs digital logic simulation for half a day, and it runs chemistry codes for the other half of the day. A benchmark suite is created consisting of two benchmarks, one of each kind. It achieves MIPS1 on the digital logic simulation benchmark and achieves

Table 4.2 The Mean to Use for Finding an Aggregate Measure over a Benchmark Suite from Measures Corresponding to Individual Benchmarks in the Suite

Measure	Valid Central Tendency for Summarized Measure over the Suite	
IPC	WAM weighted with cycles	WHM weighted with *I*-count
CPI	WAM weighted with *I*-count	WHM weighted with cycles
Speedup	WAM weighted with execution time ratios in improved system	WHM weighted with execution time ratios in the baseline system
MIPS	WAM weighted with time	WHM weighted with *I*-count
MFLOPS	WAM weighted with time	WHM weighted with FLOP count
Cache hit rate	WAM weighted with number of references to cache	WHM weighted with number of hits
Cache misses per instruction	WAM weighted with *I*-count	WHM weighted with number of misses
Branch misprediction rate per branch	WAM weighted with branch counts	WHM weighted with number of mispredictions
Normalized execution time	WAM weighted with execution times in system considered as base	WHM weighted with execution times in the system being evaluated
Transactions per minute	WAM weighted with exec times	WHM weighted with proportion of transactions for each benchmark
A/B	WAM weighted with *B*s	WHM weighted with *A*s

MIPS2 on the chemistry benchmark. The overall MIPS of the target system is the arithmetic mean of the MIPS from the two individual benchmarks and not the harmonic mean.

4.2 *Speedup*

Speedup is a very commonly used metric in the architecture community; perhaps it is the single most frequently used metric. Let us consider the example in Table 4.3.

$$\text{Total time on baseline system} = 2000 \text{ sec}$$
$$\text{Total time on enhanced system} = 1800 \text{ sec}$$

If the entire benchmark suite is run on the baseline system and enhanced system, we know that the

$$\text{Overall speedup} = 2000/1800 = 1.111$$

Now, given the individual speedups, which mean should be used to find the overall speedup? We contend that the overall speedup can be found

Table 4.3 An Example Benchmark Suite with Five Benchmarks, Their Individual Execution Times on Two Systems under Comparison, and the Individual Speedups of the Benchmarks

Benchmarks	Time on Baseline System	Time on Enhanced System	Individual Speedup
1	500	250	2
2	50	50	1
3	200	50	4
4	1000	1250	0.8
5	250	200	1.25

either by arithmetic or harmonic mean with appropriate weights. One needs to know the relative weights (with respect to execution time) of the different benchmarks on the baseline and/or enhanced system.

Weights of the benchmarks on the baseline system

$$= 500/2000, 50/2000, 200/2000, 1000/2000, 250/2000$$

Weights of the benchmarks on the enhanced system

$$= 250/1800, 50/1800, 50/1800, 1250/1800, 200/1800$$

WHM of individual speedups (weighted with time on the baseline machine)

$$= 1/(500/(2000*2) + 50/(2000*1) + 200/(2000*4)$$

$$+ 1000/(2000*0.8) + 250/(2000*1.25))$$

$$= 1/(250/2000 + 50/2000 + 50/2000 + 1250/2000 + 200/2000)$$

$$= 1/(1800/2000)$$

$$= 2000/1800$$

$$= 1.111$$

WAM of individual speedups (weighted with time on the enhanced machine)

$$= 2*250/1800 + 1*50/1800 + 4*50/1800 + 0.8*1250/1800$$

$$+ 1.25*200/1800$$

$$= (500/1800 + 50/1800 + 200/1800 + 1000/1800 + 250/1800)$$

$$= 2000/1800$$

$$= 1.111$$

Thus, if speedup of a system with respect to a baseline system is available for several programs of a benchmark suite, the WHM of the speedups for the individual benchmarks with weights corresponding to the execution times in the baseline system or the WAM of the speedups for the individual

Table 4.4 An Example in Which the Unweighted Arithmetic Mean of the Individual Speedups or the WHM Is the Correct Aggregate Speedup

Benchmarks	Time on Baseline System	Time on Enhanced System	Individual Speedup
1	200	100	2
2	100	100	1
3	400	100	4
4	80	100	0.8
5	125	100	1.25

benchmarks with weights corresponding to the execution times in the improved system can yield the overall speedup over the entire suite.

Now, consider a situation as in Table 4.4. Based on execution times, we know that the overall speedup is 905/500, which is equal to the unweighted arithmetic mean of the individual speedups. As you can see, each program had equal execution time on the enhanced machine. This is indicative of a condition in which the workload is not fixed but rather all types of workloads are equally probable on the target system. Please note that the same correct answer can be obtained if the harmonic mean of individual speedups with weights corresponding to execution times on the baseline system is used.

Next, let us consider a situation as in Table 4.5. The overall speedup is 500/380, based on the total execution times in the two systems. It can also be derived as the unweighted harmonic mean of the individual speedups. In this case, the unweighted harmonic mean is correct because the programs are equally weighted on the baseline system. It may be noted that the same correct answer can be obtained if arithmetic mean of the individual speedups, with weights corresponding to execution times on the enhanced system, is used.

One might notice that the average speedup is heavily swayed by the relative durations of the benchmarks. It is clear that the relative execution times of the benchmarks in a suite are important. However, how much thought has gone into deciding the relative durations of execution of the different benchmarks? In the CPU2000 integer benchmark suite, the baseline running times are 1400, 1400, 1100, 1800, 1000, 1800, 1300, 1800, 1100, 1900,

Table 4.5 An Example in Which the Unweighted Harmonic Mean of the Individual Speedups or the WAM Is the Correct Aggregate Speedup

Benchmarks	Time on Baseline System	Time on Enhanced System	Individual Speedup
1	100	50	2
2	100	100	1
3	100	25	4
4	100	125	0.8
5	100	80	1.25

1500, and 3000 time units for gzip, vpr, gcc, mcf, crafty, parser, eon, perlbmk, gap, vortex, bzips2, and twolf, respectively [5]. Apparently these running times were derived based on the time these programs took on a reference machine. But when metrics are aggregated assuming equal weights for each program, are we implying that twolf is thrice as important as crafty?

What mean should be used for speedups from SPEC (Standard Performance Evaluation Cooperative) benchmarks? If the aggregate number of interest is the speedup, and if the exact same SPEC benchmark suite is run in its entirety on the new system, then WHM with weights of execution times of each of the benchmarks on the baseline system should be used. This represents the condition where the target workload is exactly the same as the SPEC benchmark suite. If one argues that the relative durations of the SPEC benchmarks in the SPEC suite (as dictated by SPEC) mean nothing, the unweighted harmonic mean of speedups can be used. If one is interested in knowing the speedup of an imaginary workload in which each type of SPEC program is run for equal parts of the day on the target system, the arithmetic mean of the individual speedups should be used.

So if someone summarizes individual MIPS using unweighted harmonic mean, what does that indicate? It is a valid indicator of the overall MIPS of the suite, if every benchmark had equal number of instructions. Because either arithmetic or harmonic mean with corresponding weights is appropriate for most metrics, we can summarize the conditions under which unweighted arithmetic and harmonic means are valid for each metric. Table 4.6 presents this.

Smith uses the meaning *equal work* or equal number of floating-point operations for equal weights [1]. Under that condition, Table 4.6 does illustrate that harmonic mean is the right mean for MFLOPS. WHM with weights corresponding to number of floating-point operations or WAM with weights corresponding to the execution times of the benchmarks correctly yields the overall MFLOPS.

Ideally, the running times of benchmarks should be just enough for performance metrics to stabilize. Then, while aggregating the metrics, each program should be weighed for whatever fraction of time it will run in the user's target workload. For instance, if program 1 is a compiler, program 2 is a digital simulation, and program 3 is compression, for a user whose actual workload is digital simulation for 90% of the day, and 5% compilation and 5% compression, WAM with weights 0.05, 0.9, and 0.05 will yield a valid overall MIPS on the target workload. When one does not know the end user's actual application-mix, if the assumption is that each type of benchmark runs for an equal period of time, finding a simple (unweighted) arithmetic mean of MIPS is not an invalid approach.

4.3 Use of geometric mean

Based on the discussion in the previous sections, everything computer architects deal with can be covered by arithmetic or harmonic mean. So what is geometric mean useful for? Cragon [3] provides an example for

Table 4.6 Conditions under Which Unweighted Arithmetic and Harmonic Means
Are Valid Indicators of Overall Performance

Measure	To Summarize Measure over the Suite	
	When Is Arithmetic Mean Valid?	When Is Harmonic Mean Valid?
IPC	If equal cycles in each benchmark	If equal work (*I*-count) in each benchmark
CPI	If equal *I*-count in each benchmark	If equal cycles in each benchmark
Speedup	If equal execution times in each benchmark in the improved system	If equal execution times in each benchmark in the baseline system
MIPS	If equal times in each benchmark	If equal *I*-count in each benchmark
MFLOPS	If equal times in each benchmark	If equal FLOPS in each benchmark
Cache hit rate	If equal number of references to cache for each benchmark	If equal number of cache hits in each benchmark
Cache misses per instruction	If equal *I*-count in each benchmark	If equal number of misses in each benchmark
Branch misprediction rate per branch	If equal number of branches in each benchmark	If equal number of mispredictions in each benchmark
Normalized execution time	If equal execution times in each benchmark in the system considered as base	If equal execution times in each benchmark in the system is evaluated
Transactions per minute	If equal times in each benchmark	If equal number of transactions in each benchmark
A/B	If Bs are equal	If As are equal

which geometric mean can be used to find the mean gain per stage of a multistage amplifier, when the gains of the individual stages are given. He also illustrates that, if improvements in CPI and clock periods are given, the mean improvement for these two design changes can be found by the geometric mean. Because execution time is dependent on the product of the two metrics considered here, the mean improvement per change can be evaluated by the geometric mean. But geometric mean of performance metrics derived from component benchmarks cannot be used to summarize performance over an entire suite. A general rule is that arithmetic or harmonic means make sense when the component quantities are summed up to represent the aggregate situation. The geometric mean is meaningful when the component quantities are multiplied to represent the aggregate situation. Because execution times of component benchmarks are added to find the overall execution time, arithmetic or harmonic means should be used.

Mashey [7] presents another view for use of geometric mean. He argues that geometric mean is appropriate when metrics are distributed in a log-normal distribution as opposed to a normal distribution. A log-normal distribution is one in which the elements in the population are not distributed in a normal distribution, but their logarithms (or any base) are. He argues that speedups from programs are distributed in a log-normal fashion and, hence, that geometric mean is appropriate for speedups. However, remember that the discussions in the previous sections of this chapter are intended to find the average metric during execution of the benchmark suite. The previous sections of this chapter do not assume any distribution on how actual programs in the real world may be distributed. They do not predict the potential metric that might be obtained when some program is run on the platform of interest. The discussions were simply about computing the average while the benchmark suite was run, without assuming any particular distributions of the metrics for workloads that have not been run. A prediction of performance for another workload based on a mean of the sampled population is possible only if the programs in our benchmark suite are chosen randomly from the workload space. The advantage of a random pick is that programs will be representative of the workload space, provided a sufficiently large number of samples are taken. Often, many benchmark suites have unique and interesting programs from different parts of the workload space as opposed to randomly picked programs. Hence, it is arguable whether means from benchmark suites can be used to predict performance on actual workloads.

4.4 Summary

Performance can be summarized over a benchmark suite by using arithmetic or harmonic means with appropriate weights. If the metric of interest is obtained by dividing A by B, if A is weighed equally between the benchmarks, harmonic mean is correct; and if B is weighed equally among the component benchmarks in a suite, arithmetic mean is correct while summarizing the metric over the entire suite. If speedup of a system with respect to a baseline system is available for several programs of a benchmark suite, the WHM of the speedups for the individual benchmarks that have weights corresponding to the execution times in the baseline system can yield the overall speedup over the entire suite. The same is true for the WAM of the speedups for the individual benchmarks that have weights corresponding to the execution times in the improved system. The average performance calculated using the principles in this chapter simply represents averages over the entire suite. A prediction of performance for another workload based on a mean of the sampled population is possible only if the programs in our benchmark suite are chosen randomly from the workload space.

Acknowledgment

The feedback from Jim Smith, David Lilja, Doug Burger, John Mashey, and my students in the Laboratory of Computer Architecture helped to improve this manuscript.

References

1. Smith, J.E., Characterizing computer performance with a single number, *Communications of ACM*, 31(10), 1202, 1988.
2. Patterson and Hennessy, *Computer Architecture: The Hardware/Software Approach*, Morgan Kaufman Publishers, San Francisco, CA.
3. Cragon, H., *Computer Architecture and Implementation*, Cambridge University Press, Cambridge, U.K.
4. Lilja, D., *Measuring Computer Performance: A Practitioner's Guide*, Cambridge University Press, 2000, Cambridge, U.K.
5. The CPU2000 Results published by SPEC, online at: http://www.spec.org/cpu2000/results/cpu2000.html#SPECint.
6. John, L.K., More on finding a single number to indicate overall performance of a benchmark suite, *Computer Architecture News*, 32 (1), 3, 2004.
7. Mashey, J. R, War of the benchmark menas: Time for a truce, *Computer Architecture News*, 32 (1), 4, 2004.

Chapter Five

Statistical Techniques for Computer Performance Analysis

David J. Lilja and Joshua J. Yi

Contents

5.1 Why statistics?

Computer performance measurement experiments typically come in one of two different forms, either measurements of real systems or simulation-based studies. Each of these different types of experiments presents their own unique challenges to interpreting the resulting data. Measurement experiments, for instance, are subject to errors in the resulting data due both to noise in the system being measured and to noise in the measurement tools themselves. As a result, it is likely that the experimenter will obtain different values for a measurement each time the experiment is performed. The issues then become how to interpret these varying values and how to compare systems when there is noise in the results.

Simulation-based studies, on the other hand, typically are not affected by these types of measurement errors. If the simulator is deterministic, the output of a given simulation with the same set of inputs should be exactly the same each time the simulation is performed. One of the main difficulties with a large simulation study, though, is that it is very easy to produce a huge amount of data by varying the simulation inputs over a wide range of possible values. The problem then becomes trying to sort through this data to understand what it all means. Additionally, the optimal situation would be to minimize the number of simulations that need to be run in the first place without compromising the final conclusions that we can draw from the experiments.

This chapter will examine how statistics can help in addressing both of these issues. It will address how statistics can be used to deal with noisy measurements and how a statistical design of experiments approach can be used to sort through a large number of simulation results to aggregate the data into meaningful conclusions. In particular, it provides a tutorial explanation of how *confidence intervals* can be used to extract quantitative information from noisy data [1]. The chapter will also describe how to use a *Plackett and Burman experimental design* to help an experimenter efficiently explore a large design space in a large-scale simulation-based study [2].

5.2   *Extracting information from noisy measurements*

Experimental errors lead to *noise* in any form of measurement experiment. From the experimenter's perspective, this noise leads to *imprecision* in the measured values, making it difficult to interpret the results. It also makes it difficult to compare measurements across different systems or to determine whether or not a change to a system has produced a meaningful change in performance. It could be that what appears to be a change in performance is actually nothing more than random fluctuations in the values being measured. This section first discusses the sources of experimental errors and the concepts of accuracy, precision, and resolution of measurement tools. Confidence intervals then are introduced as a technique to quantify the precision of a set of measurements. A later section will show how to use confidence intervals to compare different sets of measurements to determine whether the changes observed in a system, or when comparing systems, are due to real effects or whether they are simply the result of measurement noise.

5.2.1 Experimental errors

There are two fundamentally different types of experimental errors: systematic errors and random errors. *Systematic errors* are the result of some sort of mistake in the experiment. For example, the experimenter may forget to reset the system to precisely the same state each time an experiment is performed, or there may be some external environmental change that affects the values that are measured. A change in the ambient temperature may cause the system's clock to change frequency slightly, for example, which would affect the values read from an interval timer that uses this clock as its time base. These systematic errors typically produce a constant or slowly changing bias in the values measured. The skill of the experimenter is the key to controlling these types of errors.

In contrast to systematic errors, the effects of *random errors* are nondeterministic and completely unpredictable. Changes in measured values that are caused by random errors are unbiased, meaning that these errors have an equal probability of either increasing or decreasing the final measured value. Random errors are inherent in the system being measured and cannot be controlled by the experimenter. They occur because of inaccuracies and limitations in the tools used to measure the desired value and because of random events that occur within the system. For instance, background operating system processes can start and stop at random times, page and cache mappings can change each time a program is executed, and so on. All of these random effects can affect the execution time of a benchmark program in unpredictable ways.

Although these events that produce random measurement errors typically cannot be controlled, they can be characterized and quantified by using appropriate statistical techniques. Before presenting these statistical techniques, it is helpful to understand how the limitations of the measuring tools themselves affect the errors observed in the measured values.

5.2.2 Accuracy, precision, and resolution

The basic metric used to quantify the performance of a computer system is usually time [3]. For instance, the time required to execute a benchmark could be measured using an interval timer built in to the computer system being tested. This time then is used as the measure of the performance of the system when executing that benchmark program. Every tool used to measure performance, however, has certain characteristics that limit the quality of the value actually measured. In fact, each time an experiment is repeated, the experimenter is likely to measure a different value.

Figure 5.1 shows an example of a histogram of a set of time measurements obtained from an interval timer on a computer system executing a given benchmark program. The horizontal axis represents the specific values that are measured in each repetition of the experiment. The vertical axis represents a count of the number of experiments in which each specific value was measured. We see that the distribution of measurements shows the

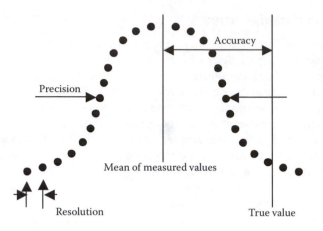

Figure 5.1 A histogram of measured values showing the accuracy, precision, and resolution of the measurement tool.

characteristic bell curve. The peak in the middle is the mean (or arithmetic average) of all of the values actually measured. We see that there also are some measured values that are larger than the mean, and an equal number of values that are smaller than the mean.

This histogram showing the distribution of the measured values demonstrates some interesting characteristics of the interval timer used to produce these measurements. The minimum distance between adjacent measured values corresponds to the *resolution* of the measuring device. This resolution is the smallest change in the phenomenon being measured that the measuring device can distinguish. In the case of the interval timer, this is the period of the timer, that is, the interval between clock ticks.

The *precision* of a measuring device is an indication of the repeatability of its measurements. Thus, the width of the distribution of measured values is a function of the precision of the measuring device. The more precise a measuring tool is, the more repeatable its results will be. Finally, the *accuracy* of the measuring tool shows how far away the mean of the values measured is from the actual or *true* value. Note that a measuring tool can be very precise without being very accurate, as suggested by Figure 5.1.

Systematic errors tend to affect the accuracy of a set of measurements. The accuracy of a specific measuring tool is hard to quantify, though, because accuracy is relative to some predefined standard. The accuracy of an interval timer, for instance, is a function of the accuracy of the oscillator used to generate the clock pulses that increment the counter that is at the core of the timer. The accuracy of this oscillator, in turn, must be compared to the standard second as defined by some appropriate standards body, such as the U.S. National Institute of Standards and Technology (NIST). The resolution of the measuring tool, on the other hand, is a characteristic of the tool itself. The resolution of the interval timer is determined by the period of the clock used to increment the timer and is, therefore, determined by the person who designed it.

Because the precision of a measuring tool is an indication of the repeatability of the values measured during an experiment, precision is most affected by random errors in the experiment. For example, the limited resolution of an interval timer introduces a random *quantization* error in the measured values. Additionally, other random events in the system will also alter the value measured each time the experiment is run, further affecting the precision of the resulting set of measurements. Although we may not be able to control the precision of the measurements, we can quantify the amount of *imprecision* using the confidence interval technique described in the next section.

5.2.3 Confidence interval for the mean

When we attempt to measure some value from a system being tested, such as the execution time of a benchmark program, we can never know for sure whether or not we have measured the true value. As discussed earlier, experimental errors lead to noise in our measurements. To try and compensate for this noise, we make multiple measurements of the same parameter. As we saw in Figure 5.1, these measurements will cover a range of values. We can use the *sample mean* value of the set of measurements as our best guess of the actual value we are trying to measure. The *sample standard deviation* of the measurements, which is denoted s, can be used to quantify the spread of the measurements around the mean. The term *sample* is applied in this situation to emphasize the fact that the mean and the standard deviation are computed from a sample of measured values and are not calculated by knowing the underlying probability distribution that produced the values measured.

The *sample variance*, which is the square of the standard deviation, is computed as follows:

$$s^2 = \frac{\sum_{i=1}^{n}(x_i - \bar{x})^2}{n-1} = \frac{n\sum_{i=1}^{n}x_i^2 - \left(\sum_{i=1}^{n}x_i\right)^2}{n(n-1)}$$

where the x_i terms are the individual measurements and n is the total number of measurements.

Although we can compare the size of the standard deviation to the mean to obtain a sense of the relative magnitude of the spread of the values measured, a confidence interval allows us to say something more precise about our measured values than using only the standard deviation. In particular, a *confidence interval* is two values—c_1 and c_2—that are centered around the mean value such that there is a 1-α probability that the real mean value is between c_1 and c_2.

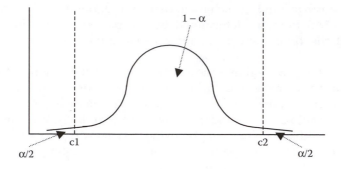
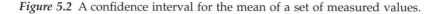

Figure 5.2 A confidence interval for the mean of a set of measured values.

As suggested in Figure 5.2, we want to find c_1 and c_2 so that

$$\Pr[c_1 \le \bar{x} \le c_2] = 1 - \alpha$$

After we find these two values, we can say with $(1-\alpha) \times 100\%$ confidence that the real mean value lies between c_1 and c_2. The value $1-\alpha$ is called the *confidence level*, and α is called the *significance level*.

To develop an equation for finding c_1 and c_2, we first normalize the measured values using the following transformation:

$$z_i = \frac{\bar{x} - x_i}{s/\sqrt{n}}$$

This transformation shifts the distribution of measured values shown in Figure 5.2 so that they are centered around 0 with a standard deviation of 1. After this normalization, the z_i values follow what is known as a *Student's t distribution* with $n - 1$ *degrees of freedom*. This distribution is very similar to a *Gaussian* (or *normal*) distribution, except that it tends to be a bit more squashed and spread out than the Gaussian distribution. In fact, as the number of degrees of freedom becomes large, the peak of the *t* distribution becomes sharper until it becomes a Gaussian distribution with mean of 0 and a standard deviation of 1. This normalization is useful for finding confidence intervals because the specific values of the *t* distribution for different degrees of freedom are easily obtained from precomputed tables [4,5,6].

Looking again at Figure 5.2, we see that c_1 and c_2 form a symmetric interval around the mean value. Thus, finding c_1 and c_2 so that the probability of the mean being between these two values is $1-\alpha$ is equivalent to finding either c_1 or c_2 such that

$$\Pr[\bar{x} < c_1] = \Pr[\bar{x} > c_2] = \frac{\alpha}{2}$$

Combining this expression with the normalization for z_i, we obtain the following expression for computing the confidence interval for the mean of the measured values:

$$c_{1,2} = \overline{x} \mp t_{1-\alpha/2;n-1} \frac{s}{\sqrt{n}},$$

where $t_{1-\alpha/2;n-1}$ is the value from the t distribution that has an area of $1-\frac{\alpha}{2}$ to the left of this value with $n-1$ degrees of freedom, s is the sample standard deviation of the measured values, and n is the total number of measurements.

Example 5.1

Consider an experiment in which you measure the execution time of a benchmark program $n = 7$ times on a given computer system. The values you measure are $x_i = \{196, 204, 202, 199, 209, 215, 213\}$, from which you compute the mean and standard deviation to be $\overline{x} = 205.4$ and $s = 7.14$. For a 90% confidence interval, $\alpha = 0.10$ so that $1-\frac{\alpha}{2} = 0.95$. The corresponding t value obtained from a precomputed table is $t_{0.95;6} = 1.943$. We then compute the 90% confidence interval to be $(c_1, c_2) = (200, 211)$. For 95% confidence, we find $t_{0.975;6} = 2.447$, which leads to a 95% confidence interval of $(c_1, c_2) = (199, 212)$.

So how do we interpret these intervals? First, the 90% confidence interval tells us that there is a 90% chance that the real mean value of the execution time of the program we measured is between 200 and 211 seconds. Note that this result further implies that there is a 10% chance that the real value is either larger than 211 or smaller than 200 due to random fluctuations in the measurements (i.e., due to the effects of experimental errors). Second, if we want to decrease the chance that the real value is outside the interval to 5%, for instance, we can use the 95% confidence interval previously computed, (199, 212). This interval must be wider than the 90% confidence interval because the only way to increase the probability that the mean is within the interval is to make the interval larger, as shown in Figure 5.3. Indeed, the only way to be 100% sure that the mean is within the interval is to push the ends of the interval out to $\pm\infty$.

Confidence intervals are useful for quantifying the spread around the mean of a set of measurements that occurs because of random errors. It is important to keep in mind, however, that the development of the preceding confidence interval formula assumes that these random errors are Gaussian-distributed. That is, the effect of the random errors on the measurements must be such that the resulting distribution of measurements follows the bell-curve shape shown in Figure 5.1. If this assumption is not true for your measurements, then the probability of the actual mean being within the computed confidence interval may not be what you expect. One approach to make confidence intervals work for any set of measured data is to *normalize* the data by averaging together several values to produce subsample means.

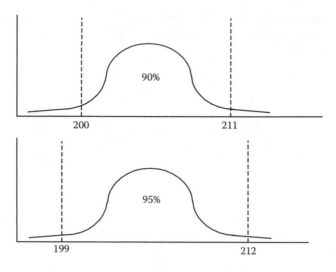

Figure 5.3 To increase the confidence that the mean value is within the computed interval, the ends of the interval must be pushed out to thereby increase the area under the curve from 90% to 95%.

We then can compute confidence intervals using these subsample means because the central limit theorem guarantees that the error in these subsample means will be Gaussian-distributed [7].

5.2.4   *Confidence intervals for proportions*

The confidence intervals described in the preceding subsection assume that the measured values are samples taken from an underlying process that produces continuous values. It is also possible to compute confidence intervals for proportions that are ratios of discrete values. For instance, assume that we take n samples of a system and find that, out of these n samples, m of them are unique in some way. For example, the value m may represent the number of packets sent over a communication network that are found to be corrupted at the receiving end, out of n total packets that are sent. The ratio $p = m/n$ then is the proportion of all packets sent that are received in error.

We can compute a confidence interval for p by recognizing that this process follows a binomial distribution with mean p and variance $s^2 = p(1 - p)/n$. Following a derivation similar to that in Section 5.2.3, the confidence interval for this proportion then is

$$c_{1,2} = p \mp z_{1-\alpha/2}\sqrt{\frac{p(1-p)}{n}}$$

In this case, the value from the t table is taken from the row with an infinite number of degrees of freedom because we can approximate a binomial distribution with a Gaussian distribution when np is sufficiently large. This tabulated value is the same as the Gaussian distribution with a mean of zero and a standard deviation of 1, which we denote $z_{1-\frac{\alpha}{2}}$.

Example 5.2

Say that we measure $n = 4591$ packets on a network and find that $m = 324$ of them are received in error. The proportion of corrupted packets then is $p = 324/4591 = 0.0706$, or approximately 7.1%. The corresponding standard deviation is $s = [0.0706(1 - 0.0706)/4591]^2 = 0.0038$. The resulting 90% confidence interval for p is $(c_1, c_2) = 0.0706 \pm 1.645(0.0038) = (0.0644, 0.0768)$. Thus, we can say with 90% confidence that the actual proportion of packets received in error on this communication network is between approximately 6.4% and 7.7%.

5.2.5 Comparing noisy measurements

The real power of confidence intervals becomes apparent when we try to compare sets of noisy measurements. In the example in Section 5.2.3, we found the mean execution time of a benchmark program to be 205.4 seconds with a 90% confidence interval of (200, 211) seconds. We now want to compare the performance of this system to another computer system when executing the same benchmark program. We run the benchmark $n_2 = 11$ times on this second system and find a mean execution time of 186 seconds with a standard deviation of 32.3 seconds. Based only on the mean values, it appears that the second system is faster than the first. However, can we be sure that this difference is *statistically significant*, or might it be due to random fluctuations in the measurements we made of each system?

Although we can never answer this question with 100% confidence, we can use the confidence interval approach to quantify the probability that the difference we see is statistically significant. In particular, we can compute a confidence interval for the difference of the two mean values. The procedure is as follows:

1. Measure n_1 values for system 1 and n_2 values for system 2, where n_1 does not necessarily have to be the same as n_2.
2. Compute the two mean values, $\bar{x}_1$ and $\bar{x}_2$.
3. Compute the difference of the mean values: $\bar{x} = \bar{x}_1 - \bar{x}_2$.
4. Compute the standard deviation for this difference of mean values, s_x (see below for details on computing s_x).
5. Compute the number of degrees of freedom that correspond to this difference of mean values, n_{df} (see below for details on computing n_{df}).

6. Using the standard deviation and number of degrees of freedom, compute a confidence interval for $\bar{x}$ as: $c_{1,2} = \bar{x} \mp t_{1-\alpha/2;n_{df}} s_x$.
7. If this interval includes 0, then we must conclude that there is no statistically significant difference between the two systems.

The combined standard deviation of this difference is the weighted sum of the individual standard deviations, where the weight is determined by the number of measurements made on each system. Thus, the standard deviation for the difference of the mean values is

$$s_x = \sqrt{\frac{s_1^2}{n_1} + \frac{s_2^2}{n_2}}$$

The formula for computing the corresponding number of degrees of freedom is

$$n_{df} = \frac{\left(\dfrac{s_1^2}{n_1} + \dfrac{s_2^2}{n_2}\right)^2}{\dfrac{\left(s_1^2/n_1\right)^2}{n_1 - 1} + \dfrac{\left(s_2^2/n_2\right)^2}{n_2 - 1}}.$$

The derivation of this formula is not at all intuitive. Furthermore, it is only an approximation that typically will not produce a whole number. Instead, the value obtained should be rounded to the nearest whole number value.

Example 5.3

We now can apply the above procedure to determine whether there is a statistically significant difference between the two systems that we measured in the previous example. Recall that for system 1, we found $\bar{x}_1 = 205.4$ seconds and $s_1 = 7.14$ for our $n_1 = 7$ measurements, and $\bar{x}_2 = 186$ seconds, $s_2 = 32.3$, and $n_2 = 11$ for system 2. From these values, we find $\bar{x} = 19.5$, $s_x = 10.1$, and $n_{df} = 11.48$, which we round to 11. For 90% confidence, we look up the tabulated value $t_{0.95;11} = 1.796$ to compute the confidence interval $(c_1, c_2) = (1.4, 37.7)$. From this confidence interval, we can conclude that there is a 90% chance that the difference in the mean values of the two systems does not include 0. That is, we are 90% certain that there is a statistically significant difference between the two systems. Of course, there still is a 10% chance that random fluctuations in our measurements caused the difference that we observe here. In this case, the difference between the two systems actually could be zero, which would mean that there is no difference between the execution times of this benchmark program on these two systems.

Example 5.4

If we want to increase our confidence that the mean is within our computed interval, we can compute a 95% confidence interval, which we find to be $(c_1, c_2) = (-2.7, 41.8)$. In this case, this larger interval includes 0. Thus, we are forced to conclude that there really is no statistically significant difference between the two sets of measured execution times at the 95% confidence level. The difference that we see could be due to measurement noise alone.

We seem to obtain two different answers from these examples, one that says that there is a statistically significant difference between the two systems, and another that says there really is no difference. So what can we conclude from these experiments? In a technical sense, we can conclude that, with 90% confidence, there appears to be a statistically significant difference. However, this difference disappears when we increase our confidence to 95%. In a practical sense, a safe conclusion would be that, yes, system 2 does appear to be faster than system 1. However, this difference is relatively small and may not be statistically significant. We also can conclude that there is a fair amount of noise in our measurements, which makes it difficult to tease out the actual differences between the two systems. To make a better comparison, we would need to run more experiments, or to obtain a better measuring tool.

5.2.6 Before-and-after comparisons

The technique for comparing noisy measurements described in the previous subsection is quite general and can be applied in any situation in which we want to compute a confidence interval for the difference between two mean values. If we know that there is a direct correspondence between pairs of measured values, though, we can apply a slight refinement to the technique given earlier. This refinement often produces tighter confidence intervals than when using the more general approach. In particular, in many types of experiments we want to see whether some change to a system produces a statistically significant change in performance. For instance, we might want to see whether adding more memory to a system actually improves the performance.

In this type of situation, which we call a *before-and-after comparison*, we can find a confidence interval for the *mean of the differences* of each pair of measured values. Let b_i be the set of n measurements made on the original (*before*) system and a_i be the set of n measurements made on the modified (*after*) system. Then the $d_i = b_i - a_i$ values are the n differences of the performance before and after the change was made. We can compute a confidence interval for the mean of these n differences, $\bar{d}$, using the same procedure described in Section 5.2.5. The resulting formula for computing the desired confidence interval is

$$(c_1, c_2) = \bar{d} \mp t_{1-\alpha/2;n-1} \frac{s_d}{\sqrt{n}}$$

where s_d is the standard deviation of the n differences, d_i. As before, if this interval includes 0, we must conclude that there is no significant difference in the before and after configurations.

Example 5.5

This type of before-and-after comparison can be used to determine whether there is a difference between two systems when executing several different benchmark programs. In this situation, each before-and-after pair has something in common that is different from the other pairs, namely, the different benchmark programs. You measure the execution times of each of five benchmark programs first on system 1 and then on system 2. You find the five execution times on system 1 to be $b_i = \{96, 89, 102, 98, 93\}$ seconds. You then execute the same five programs on system 2 and find the execution times to be $a_i = \{88, 84, 103, 90, 89\}$ seconds. The differences of each pair of before and after times are easily computed to be $d_i = \{8, 5, 1, 8, 4\}$ seconds. The mean of these differences is 4.8 seconds with a standard deviation of 3.70. For a 95% confidence interval, the necessary value from the t-table is $t_{0.975;4} = 2.777$. The resulting confidence interval then is computed to be (0.20, 9.4) seconds. Because this interval does not include 0, we conclude with 95% confidence that there is a statistically significant difference in the execution times of these two systems when executing these five benchmark programs. We also note, however, that the interval is relatively large, due to the wide variation in the measured execution times. This variation suggests that there likely are other factors affecting the execution times of the systems in addition to inherent differences between them, such as random noise in the measurements.

5.2.7 Comparing proportions

The confidence interval technique also can be used to compare two proportions, p_1 and p_2. In this case, the difference in proportions is $p = (p_1 - p_2)$ with a combined standard deviation of

$$s_p = \sqrt{\frac{p_1(1-p_1)}{n_1} + \frac{p_2(1-p_2)}{n_2}}$$

The corresponding confidence interval then is

$$(c_1, c_2) = p \mp z_{1-\alpha/2} s_p$$

As before, $z_{1-\alpha/2}$ is taken from the row in the t table with an infinite number of degrees of freedom, which is the same as the Gaussian distribution with a mean of zero and a standard deviation of 1.

Example 5.6

In the example in Section 5.2.4, we found that 324 out of 4591 packets sent on a network had errors when they were received. We now make a change to the network and find that 433 out of 7541 packets now have errors. Did this change to the network make a statistically significant difference in the error rate? To determine a confidence interval for the difference in these two proportions, we first compute $p_1 = 324/4591 = 0.0706$ and $p_2 = 433/7541 = 0.0574$. The difference of these two proportions then is $p = 0.0706 - 0.0574 = 0.0132$. The combined standard deviation for p is

$$s_p = \sqrt{\frac{0.0706(1-0.0706)}{4591} + \frac{0.0574(1-0.0574)}{7541}} = 0.0046$$

For a 90% confidence interval, $z_{1-\frac{\alpha}{2}} = 1.645$. The interval then is computed to be

$$(c_1, c_2) = 0.0132 \mp 1.645(0.0046) = (0.0056, 0.0208)$$

Because this interval does not include 0, we conclude, with 90% confidence, that this change to the system did make a statistically significant improvement in the error rate on this network.

5.3 Design of experiments

The confidence interval technique is very useful for comparing two sets of measured data. However, it is difficult to generalize it to compare more than two sets of data. Furthermore, it is not particularly useful if we want to determine the impact that each of several different input parameters has on the final measured value. A more general approach for making these types of determinations is based on the statistical *design of experiments*. The primary goal behind the design of experiments technique is to provide the most information about a system with the smallest number of experiments. A good experimental design can isolate the effects of each input variable, show the effects that interactions between input variables have on the system's output, and determine how much of the change in the system's output is due to the experimental error.

The simplest type of experimental design varies the specific values on one input while holding the others constant. Although simple, this *one-factor-at-a-time* design limits the quality of the information that can be obtained and ignores the effects of possible interactions between inputs. The most general design, and the one that produces the most detailed information, is called a *full factorial design with replication*. A full factorial design measures the response of the system when its inputs are set to all possible combinations. For experiments on real systems that are subject to the types of experimental errors previously described, this measurement process is repeated several times, or *replicated*, to allow the impact of experimental error on the output to be quantified. A mathematical technique called the *analysis of variance* (ANOVA) then can be used to extract the necessary information from the experimental data.

The basic idea behind ANOVA is to compute *sum-of-squares* terms on the measured output responses to separate the effects on the output of each input factor, the interaction between factors, and the measurement error. The effects of each input factor and the effects of their interactions can be statistically compared to the magnitude of the experimental error to determine whether the effects of each variable are statistically significant, or whether the observed response is simply due to random fluctuations in the measurements. (Further details regarding the design of experiments and the ANOVA technique applied to computer performance measurements can be found in [8,9].)

This type of analysis can be thought of as the gold-standard experimental design because it provides the experimenter with complete information about the effects of all inputs and all interactions. The problem, however, is that this full factorial ANOVA experimental design can require an unrealistically large number of experiments. For example, consider a system that has 10 inputs, each of which can take on 4 different values. Furthermore, in order to account for the experimental error, we plan to replicate each experiment 3 times. Then the total number of experiments that we would need to perform is 3×4^{10}. Thus, this experimental design would require more than 3 million separate experiments. Performing such a large number of experiments usually is prohibitively expensive, either in terms of money or in terms of the time and effort required to conduct all of the experiments. In the next section, we describe a technique that can be used to determine which parameters produce the most important bottlenecks in the performance of a system without having to measure its performance with all possible input combinations. This bottleneck analysis then can be used to simplify the problem of trying to explore a very large design space.

5.4   Design space exploration

One of the most common activities in simulation-based computer architecture research and design is *design space exploration*. For example, processor designers often need to search the potential design space to find an optimal configuration for their processor. Similarly, computer architecture researchers may want to characterize the performance of a proposed processor enhancement throughout the potential design space by using sensitivity analyses.

To explore the design space, computer architects typically use either the one-at-a-time approach or a full multifactorial design such as ANOVA. In the former approach, all N parameters are first set to their baseline values. Then, one parameter is varied from its baseline value while the values of the other parameters are fixed to their baseline values. This approach requires $N + 1$ simulations: one simulation for the baseline case and one simulation for each parameter when it is varied. One of the key weaknesses of this approach is that it does not account for potential interactions between parameters because two parameters are never both at their nonbaseline values. As a result, although this approach has a simulation cost that is approximately equal to the number of variable parameters, the inherently low quality of its results reduces its appeal.

As described in the previous section, when using a full multifactorial design such as ANOVA, the architect simulates all possible combinations of the N parameters. As a result, this approach requires b^N simulations, where b is the number of possible values for each parameter and N is the number of parameters. Although this approach quantifies the effects of each parameter and the effects of all interactions, the simulation cost can be extremely high, especially when using heavily parameterized simulators that may have several parameters for each of its many subsystems (e.g., caches, functional units, branch predictor), resulting in hundreds of variable parameters. As a result, this approach is appropriate only when the number of parameters is relatively small.

To bridge the gap between low-simulation-cost/low-detail approaches, such as the one-at-a-time approach, and high-simulation-cost/high-detail approaches, such as ANOVA, Plackett and Burman [10] introduced their statistics-based *fractional multifactorial design* in 1946, which we refer to as a *Plackett and Burman design*. The attraction of this fractional multifactorial design is that it requires a very small number of simulations while still quantifying the effects that each parameter and selected interactions have on the final outcome. In other words, the Plackett and Burman design provides information at approximately the level of ANOVA but with the simulation cost of approximately the one-at-a-time design level. More specifically, by using a Plackett and Burman design, a computer architect can quantify the effects of all single parameters in approximately N simulations or the effects of all single parameters and all two-factor interactions in approximately $2 \times N$ simulations. The latter design is called a *Plackett and Burman design with foldover* [11] and is explained in more depth in the next subsection. Figure 5.4 summarizes the differences in the simulation cost and the associated level of information of the one-at-a-time, ANOVA, and Plackett and Burman designs.

5.4.1   *Mechanics of the Plackett and Burman design*

In a Plackett and Burman design, the value of each parameter in a configuration is specified by the Plackett and Burman *design matrix*. Because Plackett and Burman designs exist only in sizes that are multiples of 4, assume that X is the next multiple of 4 that is greater than N, which is the number of parameters. The value of X must always be greater than N, such that when N is itself a multiple of 4, X must be the next multiple of 4. In the base Plackett and Burman design, that is, without foldover, there are X rows and $X - 1$ columns. With foldover, twice as many rows are needed giving a total

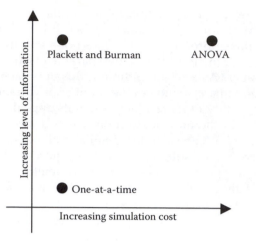

Figure 5.4 The trade-off between the simulation cost and the level of information for the one-at-a-time, ANOVA, and Plackett and Burman designs.

of $2 \times X$ rows and $X - 1$ columns. The advantage of foldover over the base Plackett and Burman design is that the effects of two-factor interactions are be filtered out from the single parameter effects.

Each row of the design matrix corresponds to a different processor configuration, and each column corresponds to the specific values for each parameter in each configuration. When there are more columns than parameters (i.e., $N < X - 1$), then the extra columns are *dummy parameters*. These dummy parameters have no effect on the simulation results and do not need to be set to any value. They exist simply to make the mathematics work properly. For most values of X, the first row of the design matrix is given in Plackett and Burman [10]. Then the next $X - 2$ rows are formed by performing a circular right shift on the preceding row. Finally, the last line of the design matrix is a row of -1s. The top half of Table 5.1 shows the Plackett and Burman design matrix when $X = 8$, which is a design matrix that can quantify the effects of up to 7 parameters. When using foldover, X additional rows (Rows 10–17 in Table 5.1) are added to the matrix. The signs in each entry of the additional rows are set to be the opposite of the corresponding entries in the original matrix. Table 5.1 shows the complete Plackett and Burman design matrix with foldover for $X = 8$.

After constructing the design matrix, but before starting the simulations, $+1$ and -1 values need to be chosen for each parameter. In a Plackett and Burman design, as in an ANOVA design, the $+1$ and -1 values represent the high and low—or on and off—values that a parameter can have. For example, the high and low values for a level-1 data cache could be 128 kilobytes (KB) and 16 KB, respectively, whereas the high and low values for speculative updates of the branch predictor could be yes and no, respectively. It is

Table 5.1 The Plackett and Burman Design Matrix, with Foldover, for X = 8

A	B	C	D	E	F	G	Exec. Time
+1	+1	+1	−1	+1	−1	−1	79
−1	+1	+1	+1	−1	+1	−1	91
−1	−1	+1	+1	+1	−1	+1	23
+1	−1	−1	+1	+1	+1	−1	24
−1	+1	−1	−1	+1	+1	+1	14
+1	−1	+1	−1	−1	+1	+1	69
+1	+1	−1	+1	−1	−1	+1	100
−1	−1	−1	−1	−1	−1	−1	39
−1	−1	−1	+1	−1	+1	+1	18
+1	−1	−1	−1	+1	−1	+1	20
+1	+1	−1	−1	−1	+1	−1	85
−1	+1	+1	−1	−1	−1	+1	38
+1	−1	+1	+1	−1	−1	−1	1
−1	+1	−1	+1	+1	−1	−1	77
−1	−1	+1	−1	+1	+1	−1	29
+1	+1	+1	+1	+1	+1	+1	1
49	264	46	39	173	45	142	

important to note that selecting high and low values that span a range of values that is too small compared to what can be reasonably expected in real systems may yield results that underestimate the effect of that parameter. On the other hand, choosing high and low values that span too large a range may overestimate the effect of that parameter on the output. Nevertheless, it is better to opt for a range that is slightly too large rather than a range that is too small to ensure that the full potential impact of each parameter is taken into account in the simulations. Ideally, though, the high and low values should be just outside of the normal (or expected) range of values.

Because the high and low values that are chosen for each parameter could be outside of the normal range of values for that parameter, the specific processor configuration found in each row of the design matrix may represent a processor configuration that is either technically infeasible or unrealistic. For instance, assume that parameter *A* in Table 5.1 corresponds to a processor's issue width, with a high value of 8-way and a low value of 2-way, and that parameter *B* corresponds to the number of entries in the reorder buffer, with a high value of 256 entries and a low value of 16 entries. For the configuration shown in row 5 of Table 5.1, the value of parameter *A* is 8-way while the value of parameter *B* is set to 16 entries. Obviously, because the reorder buffer is much too small to support an 8-way issue processor, this configuration would never be designed. However, although some of the configurations in the design matrix may not be realistic, the Plackett and Burman design still needs the results of all of these configurations because they represent the logical subset of the entire design space. Consequently, the Plackett and Burman design is architecture-independent, because its results are not dependent on the specific processor configuration.

After choosing the high and low values for each parameter, each of the configurations in the design matrix must be simulated and the corresponding output values collected. The next step is to calculate the effect that each parameter has on the variation observed in the measured output values. Note that the output value can be any metric. For example, a computer architect could calculate the effect that each of the parameters has on the execution time, branch prediction accuracy, cache miss rate, or power consumption. To calculate the effect that each parameter has on the measured output, the output value associated with each configuration is multiplied by the value (+1 or −1) of the parameter for that configuration. These products then are added across all configurations. For example, from Table 5.1, the effect of parameter G is calculated as follows:

$$\text{Effect}_G = (-1 \times 79) + (-1 \times 91) + (1 \times 23) + (-1 \times 24)$$

$$+ \cdots + (-1 \times 1) + (-1 \times 77) + (-1 \times 29) + (1 \times 1) = 142$$

After the effect of each parameter is computed, the effects can be ordered to determine their relative impacts on the variation observed in the output. It is important to note that only the magnitude of the effect is important; the sign of the effect is meaningless.

In the example in Table 5.1, it is easily seen that the parameters that have the most effect on the variation in the execution time are, in descending order: B, E, and G. From a computer architecture point of view, a parameter that has a very large effect on the variability of the execution time is a performance bottleneck because, if it is too small or not turned on, the performance will be constrained by that parameter. That is, a poor choice for a bottleneck parameter will cause the execution time to significantly increase.

Because this example uses foldover, the effect of two-factor interactions on the variation in the execution time also can be calculated. To calculate the effect of a two-factor interaction, the output value for that configuration is multiplied by the values (+1 or −1) for both of the parameters. Then, as before, the resulting products are added together. To illustrate this process, assume that the interaction of interest is AC, which is the interaction between parameters A and C. This interaction effect is calculated as follows:

$$\text{Effect}_{AC} = ((1 \times 1) \times 79) + ((-1 \times 1) \times 91) + ((-1 \times 1) \times 23)$$

$$+ \cdots + ((-1 \times -1) \times 77) + ((-1 \times 1) \times 29) + ((1 \times 1) \times 1) = -113$$

Because the resulting value of this effect is larger than the effects of all of the single parameters, except B, E, and G, the AC interaction is more of a performance bottleneck than all of the single parameters except B, E, and G. This particular result illustrates the superiority of the Plackett and Burman design with foldover when it is compared to the one-at-a-time technique. Not only does the latter technique do a poor job of quantifying the effects of the main (single) parameters, it also does not quantify the effects of any interactions between input parameters. Furthermore, the Plackett and Burman design provides this additional information while requiring only about the same number of simulations as does the one-at-a-time technique, namely, $O(N)$.

In summary, computer architects can use a Plackett and Burman design to determine the most significant performance bottlenecks in a processor and the relative ranking of all bottlenecks with respect to each other. This information can be used to pare the design space down to the most significant parameters, to thereby allow for a more efficient exploration of the design space than simulating all possible combinations of inputs.

5.4.2   Using the Plackett and Burman design to explore the design space

When trying to explore a large design space, the key problem that computer architects face is that the space of processor configurations and compiler options is enormous. Because computing resources are finite, the architect must either search a small fraction of the total configurations for many parameters, or search a large fraction of the space with a reduced number of parameters. In both cases, it is difficult for the architect to have confidence in the simulation results and its subsequent conclusions since the design space was not reduced and explored in a systematic and quantitatively based manner. However, an architect can use the Plackett and Burman design to reduce the design space with a high degree of confidence that only the most insignificant parameters were excluded.

The process for using the Plackett and Burman design to reduce the design space is very straightforward. The first step is to follow the procedure in the previous section to choose high and low values for each parameter, build the design matrix, simulate all configurations, and then calculate the effect of each parameter. Second, based on the computing resources available, and the effects of all of the parameters, select the parameters that have the largest effect on the output variable for more detailed study. Third, explore the design space of the most significant parameters using a full multifactorial design such as ANOVA. This approach, in other words, uses the Plackett and Burman design to separate the most significant parameters, which are worthy of further exploration, from the less significant ones. These less significant parameters can be set to some appropriate middle-range value as the design space for the other parameters is studied in detail. As a result, instead of simulating 2^N test cases to explore the entire design space, the computer architect can efficiently explore the design space by using the Plackett and Burman technique to first reduce the number of candidate parameters from N to N', at the cost of $2 \times N$ simulations. The reduced parameter list then can be fully explored using a full factorial ANOVA design, which has a simulation cost of $2^{N'}$ simulations.

Table 5.2 Processor Core Parameters and Their Plackett
and Burman Values (Reprinted with Permission
from [2], © 2003 IEEE)

Parameter	Low Value	High Value
Fetch Queue Entries	4	32
Branch Predictor	2-Level	Perfect
Branch MPred Penalty	10 Cycles	2 Cycles
RAS Entries	4	64
BTB Entries	16	512
BTB Assoc	2-Way	Fully Assoc
Spec Branch Update	In Commit	In Decode
Decode/Issue Width	4-Way	
ROB Entries	8	64
LSQ Entries	0.25 * ROB	1.0 * ROB
Memory Ports	1	4

To illustrate how this process works, the following example shows how the number of test cases can be reduced from 2.2 trillion (2^{41}) to 88 Plackett and Burman test cases ($X = 44$, 2×44) plus 1024 ANOVA test cases (2^{10}), for a total of 1112 test cases. Therefore, in this example, using the Plackett and Burman design to first pare the design space reduces the number of test cases by over *nine orders* of magnitude.

For this example, assume that the computer architect needs to fine-tune the processor's configuration to maximize its performance without unduly increasing its chip area. Therefore, the architect needs to accurately determine not only the most significant performance bottlenecks, but also the relative order between less significant, but still important, performance bottlenecks.

In this example, there are 41 variable parameters, which are shown in Tables 5.2, 5.3, and 5.4. Although these tables list more than 41 parameters, the variable parameters are the ones shown with high and low values. The parameters without both high and low values are static parameters; that is, they are fixed to a constant value throughout the experiments. As described in the previous section, the high and low values represent values that are just outside the normal range of values for that parameter. The normal range of values used in this example was determined by compiling a list of parameter values for several commercial processors, including the Alpha 21164 [12] and 21264 [13]; the UltraSparc I [14], II [15], and III [16]; the HP PA-RISC 8000 [17]; the PowerPC 603 [18]; and the MIPS R10000 [19].

Because there are a total of 41 variable parameters, the value of X is 44. Furthermore, in this situation, foldover is an appropriate choice because it will remove the effects of what are likely to be the most important interactions—two-factor interactions—from the effects of the single parameters. Therefore, for this example, the Plackett and Burman design matrix will have 43 columns ($X - 1$) and 88 rows ($2 \times X$). As described earlier, the

Table 5.3 Functional Unit Parameters and Their Plackett and Burman Values (Reprinted with Permission from [2], © 2003 IEEE)

Parameter	Low Value	High Value
Int ALUs	1	4
Int ALU Latency	2 Cycles	1 Cycle
Int ALU Throughput	1	
FP ALUs	1	4
FP ALU Latency	5 Cycles	1 Cycle
FP ALU Throughputs	1	
Int Mult/Div Units	1	4
Int Mult Latency	15 Cycles	2 Cycles
Int Div Latency	80 Cycles	10 Cycles
Int Mult Throughput	1	
Int Div Throughput	Equal to Int Div Latency	
FP Mult/Div Units	1	4
FP Mult Latency	5 Cycles	2 Cycles
FP Div Latency	35 Cycles	10 Cycles
FP Sqrt Latency	35 Cycles	15 Cycles
FP Mult Throughput	Equal to FP Mult Latency	
FP Div Throughput	Equal to FP Div Latency	
FP Sqrt Throughput	Equal to FP Sqrt Latency	

two extra columns are filled with dummy parameters. Consequently, the architect will need to simulate 88 different processor configurations to determine the effects of all 41 single parameters.

In Tables 5.2 and 5.4, two parameters—the number of load-store queue (LSQ) entries and the memory latency of the following blocks—are shaded in gray. For these two parameters, the high and low values cannot be chosen completely independently of the other parameters because of the mechanics of the Plackett and Burman design. The problem occurs when one of those parameters is set to one of its extreme values while the parameter it is related to is set to its opposite extreme. The resulting combination of values leads to a situation that either is infeasible or would not actually occur in a real processor. For example, if the number of LSQ entries was chosen independently of the number of reorder buffer (ROB) entries, then some of the configurations could have a 64-entry LSQ and an 8-entry ROB. Because the total number of in-flight instructions cannot exceed the number of ROB entries, however, the maximum number of filled LSQ entries will never exceed 8. Therefore, the effect of the number of LSQ entries will be artificially limited by the number of ROB entries. To avoid those types of situations, the values for all gray-shaded parameters are based on their related parameter. Although the values of these gray-shaded parameters are based on another value, they are still input parameters; basing their values on another parameter's values merely ensures that the effect of these input parameters will not be artificially limited.

Table 5.4 Processor Core Parameters and Their Plackett
and Burman Values (Reprinted with Permission
from [2], © 2003 IEEE)

Parameter	Low Value	High Value
L1 I-Cache Size	4 KB	128 KB
L1 I-Cache Assoc	1-Way	8-Way
L1 I-Cache Block Size	16 Bytes	64 Bytes
L1 I-Cache Repl Policy	Least Recently Used	
L1 I-Cache Latency	4 Cycles	1 Cycle
L1 D-Cache Size	4 KB	128 KB
L1 D-Cache Assoc	1-Way	8-Way
L1 D-Cache Block Size	16 Bytes	64 Bytes
L1 D-Cache Repl Policy	Least Recently Used	
L1 D-Cache Latency	4 Cycles	1 Cycle
L2 Cache Size	256 KB	8192 KB
L2 Cache Assoc	1-Way	8-Way
L2 Cache Block Size	64 Bytes	256 Bytes
L2 Cache Repl Policy	Least Recently Used	
L2 Cache Latency	20 Cycles	5 Cycles
Mem Latency, First	200 Cycles	50 Cycles
Mem Latency, Next	0.02 ∞ Mem Latency, First	
Mem Bandwidth	4 Bytes	32 Bytes
I-TLB Size	32 Entries	256 Entries
I-TLB Page Size	4 KB	4096 KB
I-TLB Assoc	2-Way	Fully Assoc
I-TLB Latency	80 Cycles	30 Cycles
D-TLB Size	32 Entries	256 Entries
D-TLB Page Size	Same as I-TLB Page Size	
D-TLB Assoc	2-Way	Fully Assoc
D-TLB Latency	Same as I-TLB Latency	

After choosing the high and low values for each parameter and then
creating the corresponding processor configuration files, the next step is to
run the simulations. In this example, the superscalar simulator sim-outorder
from the SimpleScalar tool suite [20] and 12 selected benchmarks from the
SPEC CPU 2000 benchmark suite [21] were used.

After calculating the effect that each parameter has on the variability in
the execution time, the parameters were ranked in descending order of effect.
This ranking provides a basis for comparison across benchmarks and ensures
that a single parameter's effect does not completely dominate the results. More
specifically, the parameter with the largest effect is given a rank of 1 while the
parameter with the second largest effect is given a rank of 2 and so on. After
ranking the parameters in descending order of effect, each parameter's rank
was averaged across all of the benchmarks. Table 5.5 shows, for each parameter, both the ranks for all benchmarks and the average rank across all benchmarks. The parameters are arranged in ascending order of their average ranks,
which corresponds to the descending order of average effects.

Table 5.5 Plackett and Burman Design Results for All Processor Parameters; Ranked by Significance and Sorted by the Sum of Ranks (Reprinted with Permission from [2], © 2003 IEEE)

Parameter	gzip	vpr-Place	vpr-Route	gcc	mesa	art	mcf	equake	ammp	parser	vortex	bzip2	twolf	Ave
ROB Entries	1	4	1	4	3	2	2	3	6	1	4	1	4	2.8
L2 Cache Latency	4	2	4	2	2	4	4	2	13	3	2	8	2	4.0
Branch Predictor	2	5	3	5	5	27	11	6	4	4	16	7	5	7.7
Int ALUs	3	7	5	8	4	29	8	9	19	6	9	2	9	9.1
L1 D-Cache Latency	7	6	7	7	12	8	14	5	40	7	5	6	6	10.0
L1 I-Cache Size	6	1	12	1	1	12	37	1	36	8	1	16	1	10.2
L2 Cache Size	9	35	2	6	21	1	1	7	2	2	6	3	43	10.6
L1 I-Cache Block Size	16	3	20	3	16	10	32	4	10	11	3	22	3	11.8
Mem Latency, First	36	25	6	9	23	3	3	8	1	5	8	5	28	12.3
LSQ Entries	12	14	9	10	13	39	10	10	17	9	7	4	10	12.6
Speculative Branch Update	8	17	23	28	7	16	39	12	8	20	22	20	17	18.2
D-TLB Size	20	28	11	23	29	13	12	11	25	14	25	11	24	18.9
L1 D-Cache Size	18	8	10	12	39	18	9	36	32	21	12	31	7	19.5
L1 I-Cache Assoc	5	40	15	29	8	34	23	28	16	17	15	9	21	20.0
FP Mult Latency	31	12	22	11	19	24	15	23	24	29	14	23	19	20.5
Memory Bandwidth	37	36	13	14	43	6	6	29	3	12	19	12	38	20.6

(continued)

Table 5.5 Plackett and Burman Design Results for All Processor Parameters; Ranked by Significance and Sorted by the Sum of Ranks (Reprinted with Permission from [2], © 2003 IEEE) (Continued)

Parameter	gzip	vpr-Place	vpr-Route	gcc	mesa	art	mcf	equake	ammp	parser	vortex	bzip2	twolf	Ave
Int ALU Latency	15	15	18	13	41	22	33	14	30	16	41	10	16	21.8
BTB Entries	10	24	19	20	9	42	31	20	22	19	20	17	34	22.1
L1 D-Cache Block Size	17	29	34	22	15	9	24	19	28	13	32	28	26	22.8
Int Div Latency	29	10	26	16	24	32	41	32	20	10	10	43	8	23.2
Int Mult/Div Units	14	20	29	31	10	23	27	24	33	36	18	26	15	23.5
L2 Cache Assoc	23	19	14	19	32	28	5	39	37	18	42	21	12	23.8
I-TLB Latency	33	18	24	18	37	30	30	16	21	32	11	29	18	24.4
Fetch Queue Entries	43	13	27	30	26	20	18	37	9	25	23	34	14	24.5
Branch MPred Penalty	11	23	42	21	6	43	20	34	11	22	39	37	23	25.5
FP ALUs	34	11	31	15	34	17	40	22	26	37	13	42	13	25.8
FP Div Latency	22	9	35	17	30	21	38	15	43	38	17	39	11	25.8
I-TLB Page Size	42	39	8	37	36	40	7	17	12	26	28	14	39	26.5
L1 D-Cache Assoc	13	38	17	34	18	41	34	33	14	15	35	15	42	26.8
I-TLB Assoc	24	27	37	25	17	31	42	13	29	30	21	33	22	27.0
L2 Cache Block Size	25	43	16	38	31	7	35	27	7	35	38	13	40	27.3
BTB Assoc	21	21	36	32	11	33	17	31	34	43	27	35	25	28.2
D-TLB Assoc	40	32	25	26	22	35	26	26	18	33	26	30	35	28.8
FP ALU Latency	32	16	38	41	38	11	22	30	23	27	30	40	29	29.0
Memory Ports	39	31	41	24	27	15	16	41	5	42	29	41	27	29.1

Dummy Parameter #2	27	42	21	39	35	14	13	35	41	28	43	18	30	29.7
FP Mult/Div Units	41	22	43	40	40	19	28	38	27	31	31	19	20	30.7
Int Mult Latency	30	41	39	36	14	26	29	21	15	41	37	32	41	30.9
FP Sqrt Latency	38	30	40	33	33	5	25	42	42	24	24	38	37	31.6
L1 I-Cache Latency	26	26	32	42	28	38	21	40	38	40	36	25	33	32.7
RAS Entries	28	33	33	27	42	25	36	25	39	39	33	36	32	32.9
Dummy Parameter #1	19	37	30	43	25	36	43	43	35	23	40	24	36	33.4

From the viewpoint of design space exploration, the key result from Table 5.5 is that, of the 41 parameters that are being evaluated (plus two dummy parameters), 10 of them are, on average, more significant than the remaining 31 parameters. This result can be clearly shown by examining the relatively large difference in the average ranks of the tenth and eleventh parameters, which are the number of LSQ entries and speculative branch update, respectively. Additionally, for each benchmark, the rank of these top 10 parameters is generally fairly low. In other words, these top 10 parameters have the most significant effect on the execution time for all benchmarks.

From these results, the computer architect can be confident that the bottom 31 parameters are insignificant compared to the top 10 parameters and can, consequently, be eliminated from the list of parameters to be studied in detail with ANOVA. Therefore, instead of performing a full ANOVA test on 41 parameters, which requires over 2.2 trillion test cases, the Plackett and Burman technique can be applied to eliminate the most insignificant parameters first. The elimination of the most insignificant parameters from further study reduces the number of test cases to a much more tractable 1024 at an additional cost of only 88 test cases.

For more information about the mechanics of ANOVA and using it to perform a detailed experimental study, read the study by David Lilja, completed in 2000 [1].

5.4.3   Other applications of the Plackett and Burman design

In addition to efficiently and accurately reducing the design space, computer architects can use the Plackett and Burman design of experiments to classify and select benchmarks and to analyze the performance of an enhancement.

To control the time required to simulate a new computer system, computer architects often select a subset of benchmarks from a benchmark suite. The potential problem with this practice is that the computer architect may not select the benchmarks in a rigorous manner, which may lead to the architect simulating a set of benchmarks that is not representative of the entire suite.

To address the problem, a computer architect can use the Plackett and Burman design to first characterize each benchmark based on the performance bottlenecks that it induces in the processor. Then, because the set of performance bottlenecks forms a unique fingerprint for that benchmark, the computer architect can cluster the benchmarks together based on the similarity of their performance bottlenecks. If two benchmarks have similar sets of performance bottlenecks, then they will be clustered together. After clustering the benchmarks into M groups, where M is the maximum number of benchmarks the architect can run, the architect needs only to choose one benchmark from each group to select a subset of benchmarks that is representative of the whole. (Chapter 9 discusses another technique for quantifying benchmark similarity.)

A Plackett and Burman design can also be used to analyze the effect that some proposed enhancement has on relieving the performance bottlenecks in a processor. By comparing the average rank of each parameter before and after the enhancement is added the processor, the computer architect can easily see which performance bottlenecks were relieved by the enhancement and which bottlenecks were exacerbated. For instance, if the average rank for a parameter increases after the enhancement is added to the processor, the enhancement mitigates the effect of that performance bottleneck. On the other hand, if the average rank decreases, then, although that enhancement may improve the processor's performance, it also exacerbates that particular performance bottleneck, which could become a limiting factor on further performance gains.

The advantage of using this approach to analyze processor enhancements is that it is not based on a single metric, such as speedup or cache miss rate, but rather on the enhancement's impact on the entire processor.

For more information about these two applications of the Plackett and Burman design, see the study done by Joshua Yi, David Lilja, and Douglas Hawkins in 2003 [2].

5.5 Summary

This chapter has demonstrated how some important statistical techniques can be used in both measurement-based and simulation-based experiments to improve the information that can be obtained from the experiments. Measurement-based experiments are subject to two types of errors: systematic errors, which are the result of some experimental mistake, and random errors, which are inherent in the system being measured and in the measurement tools themselves. Both kinds of errors produce noise in the final measurements, which causes a different value to be observed each time a measurement experiment is repeated. We showed how confidence intervals can be used to quantify the errors in the measurements and to compare sets of noisy measurements. We also showed how the design of experiment techniques can be used to efficiently search a large design space for simulation-based studies. In particular, the Plackett and Burman design is a powerful technique for finding the most important bottlenecks in a processor. Knowing these bottlenecks then allows the experimenter to substantially reduce the design space that needs to be searched by ignoring those parameters that have little impact on the final output. Taken together, the set of techniques presented in this chapter can be used to provide quantitatively defensible conclusions from computer systems performance evaluation studies.

References

1. Lilja, David J., *Measuring Computer Performance: A Practitioner's Guide*, Cambridge University Press, 2000.
2. Yi, Joshua J., Lilja, David J., and Hawkins, Douglas M., A statistically rigorous approach for improving simulation methodology, *International Symposium on High-Performance Computer Architecture* (HPCA), February 2003, 281–291.

3. Lilja, David J., *Measuring Computer Performance: A Practitioner's Guide*, Cambridge University Press, 2000, 17.
4. Dear, Keith, and Brennan, Robert, *SurfStat Statistical Tables*, University of Newcastle, June 1999, online at: http://math.uc.edu/~brycw/classes/148/tables.htm.
5. Lilja, David J., *Measuring Computer Performance: A Practitioner's Guide*, Cambridge University Press, 2000, 249–250.
6. StatSoft, Inc., *Electronic Statistics Textbook*, Tulsa, OK, 2004, online at: http://www.statsoft.com/textbook/stathome.html.
7. Lilja, David J., *Measuring Computer Performance: A Practitioner's Guide*, Cambridge University Press, 2000, 55–56.
8. Lilja, David J., *Measuring Computer Performance: A Practitioner's Guide*, Cambridge University Press, 2000, 71–77.
9. Lilja, David J., *Measuring Computer Performance: A Practitioner's Guide*, Cambridge University Press, 2000, 159–172.
10. Plackett, R., and Burman, J., The design of optimum multifactorial experiments, *Biometrika*, 33, 4, 1946, 305–325.
11. Montgomery, Douglas C., *Design and Analysis of Experiments*, 5th edition, Wiley, 2000.
12. Bannon, Peter, and Siato, Yuichi, The Alpha 21164PC microprocessor, *International Computer Conference* (COMPCON), February 1997, 20–27.
13. Kessler, Richard, The Alpha 21264 microprocessor, *IEEE Micro*, 19, 2, 1999, 24–36.
14. Tremblay, Marc, and O'Connor, J. Michael, UltraSparc I: A four-issue processor supporting multimedia, *IEEE Micro*, 16, 2, 1996, 42–50.
15. Normoyle, Kevin, Csoppenszky, Michael, Tzeng, Allan, Johnson, Timothy, Furman, Christopher, and Mostoufi, Jamshid, UltraSPARC-IIi: Expanding the boundaries of a system on a chip, *IEEE Micro*, 18, 2, 1998, 14–24.
16. Horel, Tim, and Lauterbach, Gary, UltraSPARC-III: Designing third-generation 64-bit performance, *IEEE Micro*, 19, 3, 1999, 73–85.
17. Kumar, Ashok, The HP PA-8000 RISC CPU, *IEEE Micro*, 17, 2, 1997, 27–32.
18. Song, S., Denman, Martin, and Chang, Joe, The PowerPC 604 RISC microprocessor, *IEEE Micro*, 14, 5, 1994, 8–17.
19. Yeager, Kenneth, The MIPS R10000 superscalar microprocessor, *IEEE Micro*, 16, 2, 1996, 28–40.
20. Burger, D. and Austin, T., The SimpleScalar Tool Set, Version 2.0, University of Wisconsin-Madison Computer Sciences Department Technical Report #1342, 1997.
21. Henning, J., SPEC CPU2000: Measuring CPU performance in the new millennium, *IEEE Computer*, 33, 7, 2000, 28–35.

Chapter Six

Statistical Sampling for Processor and Cache Simulation

Thomas M. Conte and Paul D. Bryan

Contents

6.1   Introduction

There are a myriad of technological alternatives that can be incorporated into a cache or processor design. Applicable to memory subsystems are cache size, associativity, and block size. For processors, these include branch handling strategies, functional unit duplication, instruction fetch, issue, completion, and retirement policies. Deciding upon which technologies to utilize among alternatives is a function of the performance each adds versus the cost each incurs. The profitability of a given design is measured through the execution of application programs and other workloads. Due to the large size of modern workloads and the greater number of available design choices, performance evaluation is a daunting task. *Trace-driven simulation* is often used to simplify this process.

Workloads or *benchmarks* may be instrumented to generate traces that contain information to measure the performance of the processor subsystem. The SPEC2000 (Standard Performance Evaluation Cooperative 2000) suite is one such benchmark suite that has been widely used to measure performance. Because these benchmarks execute for billions of instructions, an exhaustive search of the design space is time-consuming. Given the stringent time to market for processor designs, a more efficient method is required. Furthermore, storage becomes a problem because of the large amount of information contained in a trace. Statistical sampling [1,3,5,9] has been used successfully to alleviate these problems in cache simulations. In recent years it has also been extended to the simulation of processors [5,6,7].

Statistical sampling techniques involve the drawing of inferences from a sample rather than the whole, based on statistical rules. The primary goal is to make the results obtained from the sample representative of the entire workload. Thus, a critical aspect to statistical sampling is the method used to collect the samples. Sampling for caches has been thoroughly explored in the past. This chapter briefly discusses some of these methods. An accurate method for statistical trace sampling for processor simulation is then developed. The method can be used to design a sampling regimen without the need for full-trace simulations. Instead, statistical metrics are used to derive the sampling regimen and predict the accuracy of the results. When the method is tested on members of the SPEC2000 benchmarks, the maximum relative error in the predicted parallelism is less than 4%, with an average error of ±1.7% overall.

In the past, studies that have employed sampling to speed up simulation have not established error bounds around the results obtained or have used full–trace simulations to do so. Confidence intervals are necessary because they are used to establish the error that might be expected in the results. Error bounds can be obtained from the sampled simulations alone without the need for full–trace simulations. An example of validation of sampling methods for processors and the establishment of confidence intervals is included in this chapter.

6.2   *Statistical sampling*

Sampling has been defined as the process of drawing inferences about the whole population by examining only a part of that population [8]. Statisticians frequently use sampling in estimating characteristics of large populations to economize on time and resources. Sampling may be broadly classified into two types, *probability sampling* and *non-probability sampling*.

Unlike non-probability samples, probability samples are chosen by a randomized mechanism that ensures that samples are independent of subjective judgments. *Simple random sampling* is known to be one of the most accurate methods for sampling large populations. It involves a random selection of single elements from the population. However, choosing a large number of individual elements incurs a large overhead, making its application infeasible in some cases. Another less accurate, but cost-effective technique is *cluster sampling*. This technique collects contiguous groups of elements at random intervals from the population.

An element on which information is required is known as a *sampling unit*. Whereas the sampling unit for cache simulation is a memory reference, the sampling unit for a processor is a single execution cycle of the processor pipeline. The total number of sampling units from which the performance metric is drawn is called a *sample**. The larger the size of the sample, the more accurate the results. Because larger samples also mean a greater cost in time and resources, the choice of an efficient sample size is critical. A parameter in sampling theory is a numerical property of the population under test. The primary parameter for cache simulations is the *miss ratio*, whereas that for processors is the *mean instructions per cycle* (IPC).

Consider a processor running a benchmark that executes in n time cycles, $i, i + 1, i + 2, \ldots, n$, where i is a single execution cycle. For a processor, these execution cycles constitute a complete list of the sampling units or what may be termed as the *total population*. The corresponding population in cache simulations is the total set of memory references in the address trace. Simple random sampling involves random selection of sampling units from this list for inclusion in the sample. The gap between two sampling units is randomized and calculated so that the majority of the benchmark is traversed. The sampling unit immediately following each gap is included in the sample. To be able to extract single execution cycles with such precision requires simulation of the full trace, *which yields no savings in simulation cost*. Alternatively, subsets of the trace at random gaps may be extracted and executed. The execution cycles that result are then included in the sample. The random gap is calculated in the same manner as mentioned earlier. This method of sampling is essentially cluster sampling. Cluster sampling when implemented in caches has been referred to as *time sampling* [1,3,5,9].

Another technique called *stratified sampling* [15] uses prior knowledge about the elements of the population to order them into groups. Elements are then chosen from each of the groups for inclusion in the sample. This method is known as *set sampling* when applied to caches [4,9,11]. There is no known equivalent for processor sampling.

---

* Several cache trace-sampling studies refer to a cluster as a *sample*, in contrast to common statistical terminology. We will retain the statistical conventions and reserve the term *sample* for the entire set of sampling units.

6.2.1   Sample design

Sample design involves the choice of a robust (1) *sample size*, (2) *cluster size* and, (3) *number of clusters*. The accuracy of estimates for a particular sample design is primarily affected by two kinds of bias [10]: non-sampling bias and sampling bias.

Non-sampling bias arises when the population being sampled (*the study population*) is different from the *actual target* population. For example, in a full-trace cache simulation the address references at the beginning of the trace reference an empty cache. This leads to excessive misses at the start of the simulation, known as the *cold-start effect*, and can adversely affect the performance estimates. When sampling is employed, clusters are extracted from different locations in the full trace. The cache state seen by each of these clusters is not the same as in a full-trace simulation. Therefore, the cold-start effect appears at the start of every cluster. This leads to bias in the estimation of the parameter being measured. Recovering an approx-imately correct state to reduce the effect of this bias is largely an empirical sample design consideration.

The cold-start effect also affects processors. In processor sampling, the actual target population is execution cycles and the study population is trace entries. Processors maintain state in the reservation stations, functional unit pipelines, and so on. Contemporary processors have branch handling hard-ware, which also maintains considerable state.

Sampling bias is measured as the difference between the mean of the sampling distribution and the sample mean. It is a result of the sampling technique employed and the sample design. Because clusters from different locations may be selected from sample to sample, the estimates may vary across repeated samples (i.e., across repeated sampled simulations). Repeated samples yield values of means that form a distribution, known as the *sampling distribution*. Statistical theory states that, for a well-designed sample, the mean of the sampling distribution is representative of the true mean. Sampling techniques and the estimates derived from them may be prone to excessive error if the sample is not properly designed. Increasing sample size typically reduces sampling bias. In case of cluster sampling, sample size is the product of the number of clusters and cluster size. Of these two, the number of clusters should be increased to reduce sampling bias, because it constitutes the randomness in the sample design.

Sampling variability is an additional consequence due to the selection of clusters at random. The standard deviation of the sampling distribution is a measure of the variation in estimates that might be expected across samples. Making clusters internally heterogeneous (i.e., large standard deviation of the parameter within the cluster), making the cluster means homogeneous, and increasing the number of clusters are all means of reducing sampling bias and variability [8,10]. This is demonstrated for processors in Section 6.3.3.

The reduction of bias requires that the design of the sample be robust and all factors that could increase error be taken into consideration. Some of the methods that have been used to overcome or reduce the total bias are discussed in the following subsections.

6.2.2   Sampling for caches

Trace sampling has been used frequently for cache simulation studies. Two different types of sampling are possible for caches: time sampling [1,3,5,9] and set sampling [4,9,11]. Time sampling involves the extraction of time-contiguous memory references from different locations in a very long address trace. In contrast, a single set in a cache forms a member of a sample in set sampling. Therefore, the references pertaining to a set under this scheme are not necessarily time-contiguous.

6.2.2.1   Time sampling

Laha et al.   Laha, Patel, and Iyer [1] used time sampling in their experiments to show that reliable results could be obtained using trace sampling with a small number of samples. Through their work, it was shown that as little as 35 clusters of contiguous references could be used to classify the distribution of the underlying trace in all cases. Cluster sizes of 5000, 10,000, and 20,000 were used to show that a small number of clusters could correctly classify the sample traces, regardless of their length.

In this method, the *misses per instruction* (MPI) were used as a metric to determine the accuracy of trace sampling on small and large cache designs. Normally, small caches would be purged whenever a context switch is encountered. If clusters are composed of references immediately following a context switch, the behavior would be the same as in a continuous trace simulation. A continuous trace refers to the original trace that is being sampled. With this assumption, non-sampling bias is reduced by eliminating the cold-start effect.

The methodology used by Laha, et al. [1], incorporates the following steps. First, a sample size is chosen corresponding to the task interval. The average sampling interval is then calculated based upon the size of the continuous trace and the number of desired samples. Clusters of a few thousand references are collected after each sampling interval. These clusters are selected immediately following a context switch. Because of cold start after the context switch, small caches incurred very high miss rates at the beginning of the interval and generally decreased as the contents of the cache are filled. Therefore, the average value of the miss ratio was considered at the end of each sample.

For large caches, the assumption that the cache is flushed on a context switches is not valid. In cache designs that are larger than 16 kilobytes (KB), some information is almost always retained across a context switch. In this case, the non-sampling bias due to cold start cannot be eliminated as in the case of smaller caches. A new mechanism is proposed to consider only references in the trace after the point in the cluster where the cache state has been reconstructed. At the beginning of each interval, the references to the cache that cannot be determined as a hit or a miss are disregarded. Once a reference accesses a set that has been filled by previous references, it is referred to as a *primed set*. References to primed sets within the cluster are marked as significant and used for MPI calculation. References to unprimed sets are recorded as *fill references* or *unknown references* because their behavior in a full trace simulation is not known [2]. Laha, et al. [1] found that dependable estimates of miss rates were possible if significant references, or references to the primed sets, were used.

Wood et al.   Wood, Hill and Kessler [3] discussed methods to estimate the miss ratio for the *unknown (fill) references* used to warm up the cache. Whereas the fill method assumes that these references had a miss ratio equal to the overall miss ratio, Wood, Hill, and Kessler showed that the miss ratio of such references is in fact higher than the overall miss ratio.

This study models each block frame in the cache in terms of generations. A block frame is a part of the cache set capable of holding a single block. Each generation is composed of a live time and a dead time. A block frame is said to be live if the next reference to that frame is a hit, and dead if the next reference to it is a miss. A generation therefore starts after a miss occurs and ends when the next miss occurs. The miss that ends the generation is included in the generation, whereas the miss that starts it is not. The miss ratio at any instant in time during a simulation is the fraction of block frames that are dead at that instant.

The probability that a block frame is dead at any instant in time is the fraction of the generation time during which the block is dead. Assuming that the live and dead times for the block frames are identically distributed for all the block frames in the cache, the miss ratio μ_{long} is given by:

$$\mu_{\text{long}} = \frac{E[D_j]}{E[G_j]} , \qquad\qquad (6.1)$$

where $E[D_j]$ = Expected dead time in generation j, and $E[G_j]$ = Expected generation time for generation j.

Because the distributions of the live and dead times are not known, the two times can be calculated as means of the respective times computed throughout the trace. When sampling is employed, these are computed using only the sampled references. The live and dead times for each block frame are counted in terms of the number of references to that block frame. Equation (6.1), is valid only when every block in the cache is referenced at least once. Thus, only when large clusters are used can this technique for estimation be employed. This miss ratio (μ_{long}) computed in this method is the miss ratio for the unknown fill references.

For short traces it may not be possible for every block frame to be referenced at least once, making the preceding method inaccurate. Wood et al. suggest a difference procedure for estimating the miss ratio of *unknown references* for short traces. This miss ratio is based on the assumption that block frames not referenced are dead. For a cache with S sets and associativity of A, the total number of block frames is SA. If U is the number of unknown references then, $(SA - U)$ is the number of block frames that are never referenced by the cluster. Therefore,

$$\mu_{last} = \frac{\max\left(0, SA \times \frac{E[D_j]}{E[G_j]} - (SA - U)\right)}{U} \tag{6.2}$$

It is possible that not all live block frames are referenced by a small cluster. Therefore, the number of dead blocks may out-number live ones, so there is a max function with 0 in Equation (6.2). Another metric, μ_{split} is the arithmetic mean of μ_{long} and μ_{last}. μ_{tepid} simply assumes that exactly half of the block frames are dead; that is, 50% of the unknown references are misses. Therefore, μ_{tepid} is defined as 0.5. Empirical results show μ_{split} and μ_{tepid} to be the best estimators. The μ_{tepid} metric may be preferred over μ_{split} because it requires no computation.

Fu and Patel. Work by Fu and Patel [9] suggest that the miss ratio alone is not adequate. Other models estimate only the fraction of fill references that are misses and therefore only calculate the cache miss ratio. Although sufficient for some studies, this is insufficient when more detailed simulation of cache events is required. In the case of simulation of cache miss events when other system components are included, such as a multi-processor, each fill reference must be identified as a hit or a miss. Because of the cold-start effect larger caches result in a large number of fill references at the beginning of a cluster.

Fu and Patel propose a new metric for identifying fill references based upon the *miss distance*, which is the number of references between misses including the first miss. This metric is similar to the generation time used by Wood et al. [3] The results were validated by comparing the distributions of miss distance for the sampled and continuous traces. Using the miss distance, the state of fill references were predicted based on the miss history of the reference stream. A set of approximately 40 samples are selected as before, where each sample is split into a priming and evaluation interval. During the priming period, references are used to warm up the cache by generating sets of filled cache locations. By warming up the cache, the number of fill references is reduced during the evaluation interval. In the evaluation interval, each fill reference is predicted as a hit or a miss using the miss-distance history and the cache contents.

The following steps in the algorithm are applied to each sample. First simulate the priming interval of the sample and apply the history table. This history table is a small list of the most recent miss distances. During the priming interval, if a miss occurs, then the miss distance is calculated and stored in the history table. If a fill references is encountered in the priming interval, it is ignored. Fills found in the evaluation period are predicted according to the following criteria: If the history table is empty, then no misses have been recorded, then predict a hit. Otherwise, if the history table is not empty, and the distance is within the range of distances recorded in the history table, then predict a miss. If a prediction cannot be made based on the history table, then the contents of the cache are searched. If adjacent sets to the set being filled contain the block being loaded, then predict a hit. For all other cases predict a miss.

For this experiment, the history table was very small and only contained three distances. However, increasing the size of the table did not yield any performance gain. This method predicted accurate mean and standard deviation to the miss distance behaviors of the continuous trace. By searching the contents of the cache, this study assumes that the cache blocks are not replaced due to a large cache. When simulating with a smaller cache, this assumption does not hold.

Conte, et al.   This study [17] extended for sampling single pass methods that can collect an entire cache design space in one run. In so doing, [17] removes all non-sampling bias by keeping the caches warm between clusters using an LRU (last recently used) stack.

6.2.2.2   Set sampling

A cache that can hold C blocks, and has associativity A can be divided into C/A sets (i.e., each set contains A block frames). The set sampling method varies from the time-based techniques mentioned earlier, because in this approach the sets in the cache are sampled rather than the workload. The sets for inclusion in the sample may either be selected at random or by using information about the parameters of the caches. The method employed in Liu and Peir's "Cache Sampling by Sets" [16] consists of two phases. The first phase uses a partial run of the workload on the whole cache to obtain information about the behavior of each set in the cache. Based on this information, certain sets are selected for inclusion in the sample. The actual simulation is done in the second phase using only the sets in the sample. Another interesting method is that suggested by Kessler et al. [4]. Referred to as the *constant-bits* method, it can be used to simulate a hierarchy of multimegabyte caches. It can also be used to simulate multiple caches in a single simulation. Both of these methods are explained in the subsections that follow.

Liu and Peir.   Liu and Peir characterize each set by a metric called *weighted miss*, as illustrated in Figure 6.1 [16]. The sampling procedure is

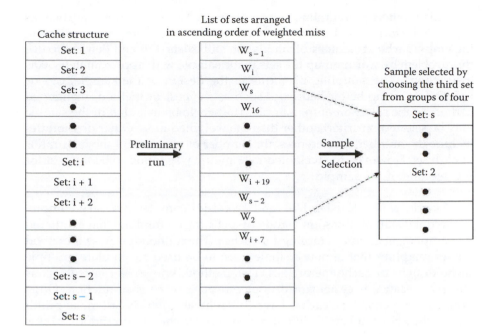

Figure 6.1 Weighted miss set selection.

initiated with a preliminary run using a subset of the workload. Liu and Peir used 15 million address references for this purpose. Let μ_{prel} be the miss ratio of the cache under study for this phase of the sampling procedure. Let μ_i be the miss ratio of the *i*th set in the cache due to the references r_i to the set. The weighted miss, W_i, for set *i* is given by

$$W_i = (\mu_i - \mu_{pred}) \times r_i \tag{6.3}$$

In words, the weighted miss of a set is the number of misses that may be attributed to the references to that set. After the preliminary run, the weighted miss metric is computed for every set in the cache. The sets are arranged in ascending order of W_i. The list of sets is then divided into equal sized groups. One set from each group is chosen for inclusion in the sample according to some heuristic. One heuristic is to choose the *p*th set from each group. Other heuristics that were seen to perform well were the *median* and *best-fit* methods. Under the median heuristic, the set with the median weighted miss value in the group is selected. With the best-fit method, the set whose weighted miss value is closest to the average weighted miss of the group is chosen. The second phase of the procedure simulates the sets chosen during the first phase. The complete workload is simulated on these sets. The miss ratio is then computed as the ratio of misses to the references to the sets in the sample.

This method of sampling does not suffer from non-sampling bias as much as the time-based techniques. However, the non-sampling bias due to the empty cache at the start of simulation still exists. Liu and Peir overcome this problem by warming up the sets in the sample with approximately 500K instructions. The sampling bias, due to the design of the sample, can be reduced by using better heuristics such as the best-fit method, mentioned earlier, for the selection of the sample. The sets to be included in the sample may be selected on criteria other than the weighted miss. Other possibilities include the number of references, the number of misses, and the miss ratios of each set. However, the weighted miss metric was found to be the best for set selection of the sample.

Kessler et al.   Kessler, Hill, and Wood [4] completed a very comprehensive and statistically sound study of cache trace sampling, that compared set sampling and time sampling for caches. The authors propose a method for set sampling that allows a single trace to be used to simulate multiple cache designs or cache hierarchies. This method, known as *constant-bits*, is effective because it systematically selects cache sets for simulation. Often primary and secondary caches have different sets, which can make multiple-level cache simulation difficult when sets are chosen at random. MPI is used to gauge cache performance in this study. According to Kessler et al. the method of MPI calculation is very important when utilizing set sampling. Two possible ways of calculating MPI are with the *sampled-instructions* and *fraction-instructions* methods. Every instruction includes the instruction fetch as well as the data references for that instruction.

Consider a sample S with n sets from a cache containing a total of s sets. The misses m_i and instructions fetches $instr_i$ for each set i in the sample are determined.

Under the sampled-instructions method, the MPI of the sample is calculated by normalizing the number of misses by the instruction fetches:

$$MPI_{Sample} = \frac{\sum_{i=1}^{n} m_i}{\sum_{i=1}^{n} instr_i} \qquad (6.4)$$

In the fraction-instructions method, the MPI of the sample is calculated by normalizing the number of misses by the fraction of sampled sets times all instruction fetches.

$$MPI_{Sample} = \frac{\sum_{i=1}^{n} m_i}{\frac{n}{s} \sum_{i=1}^{s} instr_i} \qquad (6.5)$$

These two techniques for MPI calculation were then used in conjunction with the constant-bits method. The constant-bits method applies a filter to

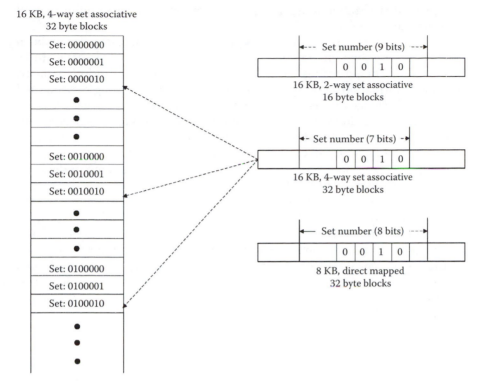

Figure 6.2 Constant-bits set selection.

only pass through address references that access the same set. Those references that pass the filter are applied to the cache design, and the sets referenced within the cache are the sample. The method is illustrated in Figure 6.2. If p bits in the set selection portion of the address are used to filter the address references, $(1/2^p)$th of the cache sets in each cache are included in the sample. In the experiments conducted by Kessler et al. the fraction-instructions method proved to be more accurate in the calculation of the sample MPI.

This method supports the simulation of multiple cache designs and cache hierarchies. The trace of address references to a secondary cache consists of the references that miss in the primary cache. When sets are selected at random it is difficult to simulate a hierarchy of caches. The misses generated from a randomly sampled primary cache when applied to a randomly sampled secondary cache do not provide reliable estimates. The constant-bits method does not encounter this problem and may be conveniently used to simulate a hierarchy of caches. However, the method implies that the samples are systematically chosen via address selection. This is a disadvantage to other methods where the sets are chosen randomly. In the case of a workload that exhibits a regular pattern, the systematic sampling could produce flawed results.

To summarize, there are two widely accepted sampling methods in caches. Set sampling chooses sets from the cache and considers these to be representative of the entire cache. The choice of sets may be random or based on some information about the sets in the cache (e.g., sampling by weighted misses). The choice of sets may also be a consequence of information available in the trace as in the constant-bits method. Set sampling has been found to provide accurate estimates at low simulation cost [4]. However, it fails to capture time-dependent behavior (such as the effects of prefetching). Although set sampling reduces the time required for simulation, it does not solve the trace storage problem. If many different caches are to be simulated the full trace needs to be stored. Time sampling, on the other hand, requires the storage of only the sampled portion of the trace. It can also capture time-dependent behavior. The drawback of time sampling is the non-sampling bias due to the cold-start effect. Many different techniques have been employed to overcome this bias. Most of these methods require additional references in each cluster thus lengthening simulation. The decision as to which method to use depends on the resources available and the desired nature of the simulation.

6.2.3   *Trace sampling for processors*

The sampled unit of information for processor simulations is not the instructions in the trace, but rather the execution cycles during a processor simulation. A metric that may be measured for each execution cycle is the IPC. Because IPC varies between benchmarks, the *relative error*, RE(IPC), may be used to validate results. The relative error is given by,

$$RE(IPC) = \frac{\mu_{IPC}^{true} - \mu_{IPC}^{sample}}{\mu_{IPC}^{true}} \qquad (6.6)$$

where μ_{IPC}^{true} is the true population mean IPC, and μ_{IPC}^{sample} is the sample mean IPC. RE(IPC) relies on μ_{IPC}^{true} from a full-trace simulations of each test benchmark. Reduction in sampling bias, sampling variability, and determination of error bounds do not require μ_{IPC}^{true}.

Conte.   The earliest study of trace-sampled processor simulation used a systematic sampling method [5]. For state repair, a strategy similar to that used for caches by Laha et al. was used. The method used 40 contiguous clusters of sizes either 10,000 or 20,000 instructions each at regular intervals. Results for a highly parallel microarchitecture with unlimited functional units showed a maximum relative error of 13% between the sampled parallelism and the actual value.

Poursepanj.   In a similar study [6], performance modeling of the PowerPC 603 microprocessor employed a method using 1 million instructions in 200 clusters of 5000 instructions each. The geometric mean of the parallelism for the SPECint92 benchmarks was within 2% of the actual value. However, the error for individual benchmarks varied as much as 13%. As with [5], the error was described using a comparison between the sampled and the full-trace simulations.

Lauterbach.   Lauterbach's study discussed an iterative sampling-verification-resampling method [7]. The sampling method used consists of extracting 100 clusters of 100,000 instructions each, at random intervals. Quick checks involving instruction frequencies, basic-block densities, and cache statistics are done to investigate the validity of the sample. The checks are done against the full trace for the benchmark. In some cases the sampled trace is not representative of the full trace. When this occurs, additional clusters are collected until the required criterion is reached. Final validation of the sampled trace compares the execution performance of the sampled trace with that of the full trace. This study simulates both the cache and the processor. The state of the cache at a new cluster is stored along with the instructions of the cluster. This state is loaded in before the beginning of the cluster during the sampled trace simulation. This reduces the influence of the cold-start effect in the cache subsystem on the processor simulation. The need to collect cache statistics makes a full-trace simulation necessary. The process of collecting the trace can therefore be time consuming. The full-trace simulation is also required to validate the sampled trace and determine error bounds.

Conte et al.   Menezes presents a thorough simulation regimen for processor simulation that allows for the calculation of confidence intervals [19]. (This chapter parallels Menezes's approach and presents updated techniques for simulations employing caches.)

Haskins and Skadron.   The study by Haskins and Skadron discusses techniques to reduce execution times in sampled simulations by devising methods that approximate the full warm-up method [18]. In full warm-up, every instruction that is skipped is applied to the state of the system in functional mode. After instructions have been skipped, then normal cycle timing accurate simulation resumes. Functional warming was originally described for cache simulation in Conte et al. [17], and later used by Wenish et al. [14] in processor simulation. Although accurate, the full warm-up method is very heavy-handed. This paper proposes methods to determine the number of instructions prior to a cluster for warm-up rather than all instructions between clusters. The two methods used to determine the number of precluster instructions for warm-up are *Minimal Subset Evaluation* and *Memory Reference Reuse Latency*. For more information, please refer to Haskins and Skadron [18].

6.3   An example

A solid body of work exists for the application of trace sampling for cache simulations. This is, however, not true for processor simulations. The remainder of this chapter demonstrates how sampling techniques can be applied to processors. The problems unique to trace sampling in processor simulations are discussed. An accurate method to alleviate non-sampling and sampling bias using empirical results is presented. Also shown is a method to calculate error bounds for results obtained using sampling techniques. These bounds can be obtained without full simulations using the sampling results alone.

Where previous studies have tried to reduce all bias as a whole and make a prescription for *all* trace-sampled processor simulation, this study separates bias into its non-sampling and sampling components. It develops techniques for reducing non-sampling bias. Reduction in sampling bias is achieved using well-known techniques of sampling design [8,10].

As the first step in the sampling process, clusters of instructions are obtained at random intervals and potentially written to a disk file. The choice of clusters at random satisfies the conditions of probability sampling. The clusters of instructions are then simulated to obtain clusters of execution cycles. The fixed number of instructions in a cluster yields a variable number of execution cycles. Statistics are ultimately calculated from these execution cycles. The number of execution cycles that would be obtained on the execution of a sampled instruction trace, N_E^{sample}, is given by,

$$N_E^{sample} = \frac{N_I^{cluster} \times N_{cluster}}{\mu_{IPC}} \tag{6.7}$$

where, $N_I^{cluster}$ is the number of instructions in a cluster, $N_{cluster}$ is the number of clusters, and μ_{IPC} is the mean IPC. The term *cluster* is used interchangeably for the group of instructions that yield a set of contiguous execution cycles, and for the set of execution cycles themselves.

6.3.1   The processor model

A highly parallel processor model is used in this study to develop a robust non-sampling bias reduction technique and to test the method for sample design. The model is an execution-driven simulator based on SimpleScalar [20], which models the MIPS R10000 processor. Unlike trace-driven simulations, the processor model fetches instructions from a compiled binary. The front end of the processor includes a four-way 64 KB instruction cache with a 64-byte line size. The superscalar core can fetch and dispatch eight instructions per cycle, and can issue and retire four instructions per cycle. The model also includes eight universal function units that are fully pipelined. The maximum number of in flight instructions is 64. The issue queue size is 32, and there is a load store queue of 64 elements. The pipeline depth is seven stages. The minimum branch miss-prediction penalty is five cycles. The processor frequency is assumed to be 4 gigahertz (GHz).

The model also includes both a functional and a timing simulator. The functional simulator is useful in a variety of different ways. First, the functional simulator is used to validate the results of the timing simulator. If the timing simulator attempts to commit a wrong value, the functional simulator will assert an error. However, in the context of trace sampling, the functional simulator has additional uses. As instructions in the dynamic instruction stream are skipped, the functional simulator still contains the valid state of the architectural register file. When a cluster is entered, the values of the registers contained in the functional simulator are copied to the timing simulator. In this manner, instructions that consume values from instructions that were skipped in between clusters will still yield correct results.

The simulator also incorporates a sophisticated memory hierarchy. The first-level data cache is a four-way 32 KB cache with 64-byte line size. The second-level unified cache is an eight-way 1 MB cache with a 64-byte line size. There are also two buses that are used to emulate the bus contention and transfer delay between the levels of memory. The first-level bus is shared between the first-level data cache and the instruction cache, and connects the first-level caches and the unified second level. The first-level bus has a width of 32 bytes and operates at 2 GHz. The second-level bus connects the second-level cache and the memory. This bus is 16 bytes and operates at 1 GHz. Table 6.1 shows some of the processor model design parameters.

Highly accurate branch prediction and speculative execution are generally accepted as essential for high superscalar performance. In the spirit of the other high-performance design parameters, a hardware predictor with high prediction accuracy is incorporated. The branch predictor used is a 64K entry Gshare predictor [12] with a 1024 entry return address stack. In addition, the processor is able to use the results of the predictor to speculatively execute beyond eight branches (for comparison, the PowerPC 604 can speculate beyond two branches [13]).

The standard performance metric for superscalar processors is the IPC, measured as the number of instructions retired per execution cycle. IPC is

Table 6.1 Simulator parameters

Issue rate: 4 instructions/cycle
Scheduling: out-of-order, reorder buffer based
Branch handling: G-share predictor
Branch speculation degree: 8 branches ahead

Table 6.2 Studied benchmark population

benchmark	True IPC
gcc	0.87314
mcf	0.20854
parser	1.07389
perl	1.28956
vpr	1.18062
vortex	0.92672
twolf	0.97398
art	0.77980
ammp	0.24811

ultimately limited by the issue rate of the processor, because flow out of the processor cannot exceed the flow in.

6.3.2 Reduction of non-sampling bias

Experiments were conducted using the SPEC2000 benchmarks. Integer benchmarks used include gcc, mcf, parser, perl, vortex, vpr, and twolf. Floating-point benchmarks used include ammp and art. Table 6.2 shows the true$_{IPC}$ of each benchmark simulated during experimentation. The first 6 billion instructions from each benchmark were simulated at the cycle level to serve as a baseline for comparison to the various sampling techniques.

Using these values, a study of the non-sampling bias for the processor model was conducted. The sampling parameters were chosen by performing a search of the design space. Simulations were performed by varying either the number of clusters or the cluster size. Figures 6.3 and 6.4 show the results of the design space search using two different warm-up methods. These methods are discussed later. From this data, cluster sizes of 1000 to 10,000 instructions were selected with a 1000-instruction step size. The cluster count was made large enough so that it did not contribute considerably to the error, but small enough to minimize the instructions to be executed. For these experiments, 2000 clusters were chosen. The first cluster was selected as the first $N_I^{cluster}$ instructions from the trace. After the first cluster, a number was randomly chosen to determine the number of skipped instructions between sampling units. A maximum interval was calculated for each cluster size such that all clusters would be selected in a uniform distribution of the first 6 billion instructions in the dynamic stream. In order to keep the sampling bias within the clusters constant, a single random seed was used for all simulations.

Using this framework a number of different techniques were used to analyze and effectively remove the amount of non-sampling bias from the sample. As discussed earlier, non-sampling bias is caused by the loss of state information during skipped periods. After a cluster is executed and instructions are skipped, the potential for state loss is high and will likely affect the performance

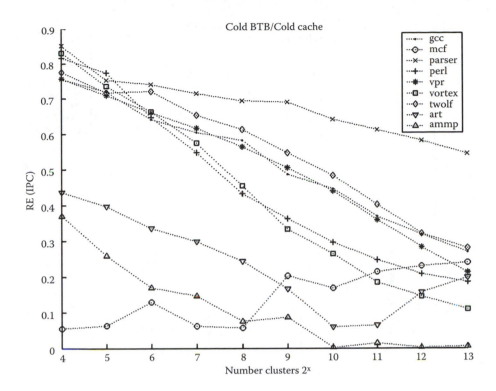

Figure 6.3 Cold BTB/cold cache.

of the next cluster. State in a processor is kept in a number of areas including: the scheduling queues, the reorder buffer, the functional unit pipelines, the branch handling target buffer (the BTB), instruction caches, data caches, load/ store queues, and control transfer instruction queues.

The following methods were simulated to analyze the affect of cold start and to remove the bias that negatively impacts sampling performance. In the *no warm-up* method, no state repair techniques were used when executing clusters. After the execution of a cluster, the caches and BTB were left cold, or stale. That is, when skipping instructions in between clusters, the BTB and the contents of the caches were at the final state of the previous cluster when execution of the next cluster began. In the *fixed warm-up* method, the state repair technique consisted of a fixed number of instructions upon entering a cluster. Using this method, no statistics were collected until the warm-up period in the cluster had finished. A certain number of instructions were used to help restore the state of the system before counting the instructions as significant for IPC statistics. For each of the cluster sizes, fixed warm-up percentages were chosen between 10% and 90%.

Functional warming techniques were also applied during experimentation. *Functional warming* refers to the warming of state in between clusters while

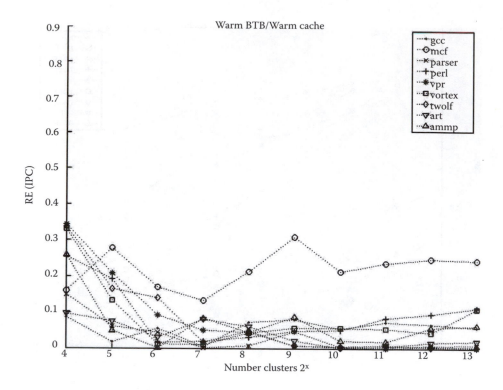

Figure 6.4 Warm BTB/warm cache.

instructions are being skipped, and was applied to the BTB and caches. In the *stale BTB–warm cache* method, the state of the L1, L2, and instruction caches were warmed in between clusters, but the BTB was left stale. In the *warm BTB–stale cache* method, the state of the BTB was warmed in between clusters, but the state of the caches were left stale. Finally, in the *warm BTB–warm cache* method both the caches and the BTB were warmed in between clusters.

A study of the non-sampling bias for the processor model is shown in Figures 6.5 though 6.11. This data shows the absolute value of the relative error between a complete run of the benchmark and the sampled run.

The results of no warm-up are shown in Figure 6.5. In this method, the state of the BTB and caches at the end of the cluster was left unchanged at the execution of the next cluster. This method assumes that no substantial changes in BTB or the caches occurred while skipping instructions, which is obviously untrue for most applications. In this method no attempt is made to mitigate the affects of cold start before using instructions within a cluster for the calculation of IPC. However, even with no state repair some interesting trends are noticed. In Figure 6.5, the relative error generally decreases as the cluster size increases. This correlates to sampling theory, which states that error should decrease as the number of samples,

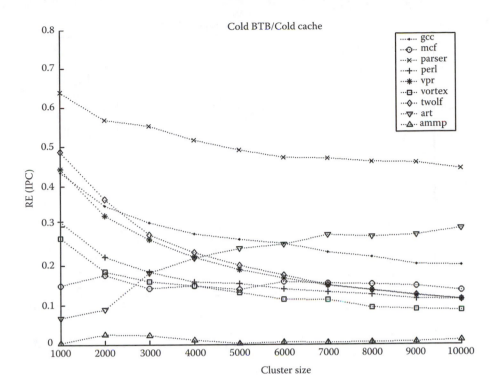

Figure 6.5 Cold BTB/cold cache.

or the size of those samples, increase. In Figure 6.5, all of the integer benchmarks exhibited this behavior. For these benchmarks, the average decrease in error was 14.3% as the cluster size increased. The twolf benchmark showed the highest accuracy gain from 48.6% to 11.3%. The smallest gain was mcf, which went from 14.9% error to 13.6%. The most striking observation from this method was the behavior exhibited by the floating-point benchmarks, which showed an increase in relative error as the cluster size increased. The art benchmark lost accuracy by 22.6% whereas ammp decreased by 0.7% at the highest cluster size. The average relative error for no warm-up for a 10,000 cluster size is 16.8%.

In Figures 6.6, 6.7, and 6.8 the results of the *fixed warm-up* method are presented for fixed warm-up percentages of 10%, 50%, and 90%, respectively. In the fixed warm-up method, cold start was addressed by using instructions within the cluster itself to warm the state of the processor before recording statistics for IPC calculation. In this method, all information during the warming period is discarded. The state of the caches and BTB are left stale just as in the no warm-up method. Figure 6.6 shows the affects of using 10% of the cluster size as a warming period. This figure looks very similar to the results presented in no warm-up. As in no warm-up the relative error

Performance Evaluation and Benchmarking

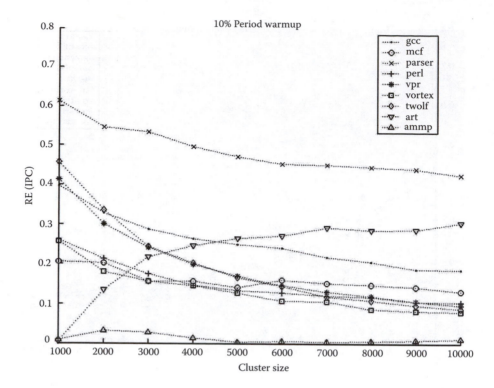

Figure 6.6 Ten percent period warm-up.

decreases as the cluster size increases for gcc, mcf, parser, perl, cpr, vortex, twolf, and ammp. In addition, the accuracy of the relative error increased for a given cluster size, as the fixed percentage of instructions for warm-up also increased, as shown in Figures 6.7 and 6.8. Although it is hard to see from the graphs, the relative error decreases marginally. When compared to no warm-up, inaccuracy gain was achieved through all of the integer benchmarks. The perl benchmark saw the highest gain at 6.7%. The lowest gain from no warm-up among the various integer benchmarks was parser at 0.9%. As in no warm-up, the art benchmark experienced a performance loss under fixed warm-up as the cluster sizes increased. The average relative error for fixed warm-up with a 90% warm-up period is 13.9%.

In Figures 6.9 through 6.11 the results of functional warming when applied to the BTB and cache structures, are presented. In these methods, the skipped instructions in between clusters were used to reduce the cold-start effect. All branches in between clusters were applied to the branch predictor in the case of BTB. For the caches, the data from loads and stores were used to keep the state of the caches consistent as they would be under the full timing simulation. In the stale BTB–warm cache method, the average relative error fell from 13.9% in fixed warm-up to 2.8%. In this mode, all benchmarks, excluding ammp, saw a remarkable drop in relative error

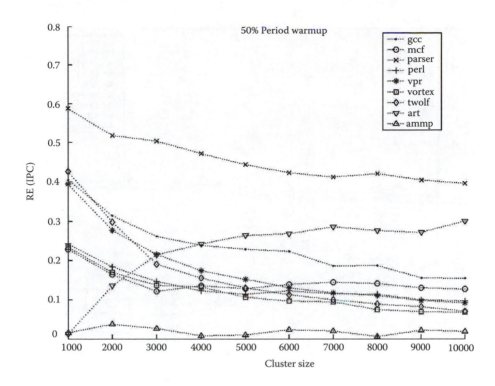

Figure 6.7 Fifty percent period warm-up.

compared to the previous methods. The parser benchmark error was reduced from 44.3% in no warm-up to 4.7%. The twolf benchmark relative error was reduced from 37.3% to 1.2%, and mcf was reduced from 13.6% to 3.4%. The importance of the BTB relative to the caches is shown in Figures 6.9 and 6-10. Even when the BTB is functionally warmed in between clusters, the penalties of all incurring cache misses at the beginning of the next cluster completely dominates the performance. As evident in the average relative errors for these two methods, the affect of warming the BTB is much less significant when compared to warming the caches. The warm BTB–stale cache method had an average relative error of 16.1%, very similar to stale BTB–stale cache. The most accurate of all methods was the warm BTB–warm cache method which had an average relative error of 1.5%.

6.3.3 Reduction in sampling bias and variability

It is accepted in sampling theory that bias exists in every sample because of the random nature of the sample. It is possible to predict the extent of the error caused by this bias. The *standard error* of the statistic under consideration is used to measure the precision of the sample results (i.e., the error bounds) [8]. Standard error is a measure of the expected variation between

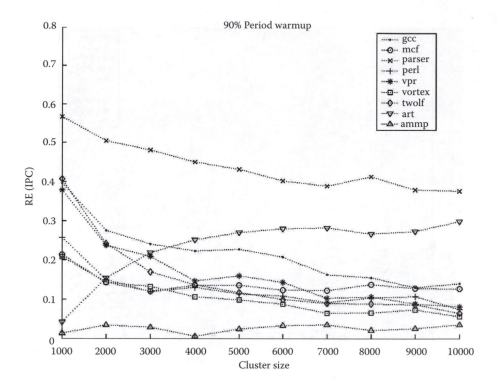

Figure 6.8 Ninety percent period warm-up.

repeated sampled simulations using a particular regimen. These repeated simulations yield mean results that form a distribution. The standard error is defined as the standard deviation of this distribution. Its use is based on the principle that the mean results of all simulations for a particular regimen are normally distributed, regardless of whether or not the parameter is normally distributed within the population. Based on this principle, the properties of the normal distribution can be used to derive the error bounds for the estimate obtained from a simulation.

It is not cost-effective to perform repeated sampled simulations to measure the standard error. Sampling theory allows the estimation of the standard error from a single simulation. This is termed as the *estimated standard error* and is denoted by S_{IPC}. This method of measurement and the results obtained from it are used in the rest of this section. The standard deviation for a cluster sampling design is given by,

$$S_{IPC} = \sqrt{\frac{\sum_{i=1}^{N_{cluster}} \left(\mu_{IPC}^i - \mu_{IPC}^{sample}\right)^2}{N_{cluster} - 1}}, \tag{6.8}$$

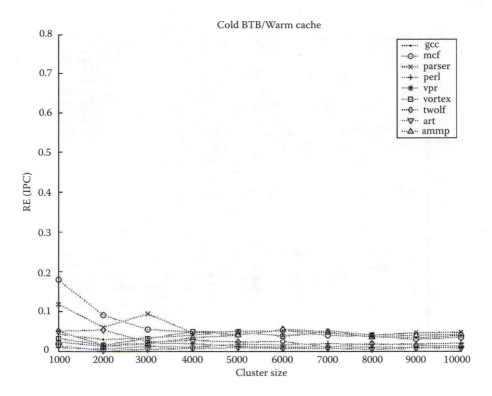

Figure 6.9 Cold BTB/warm cache.

where μ_{IPC}^{i} is the mean IPC for the ith cluster in the sample. The estimated standard error can then be calculated from the standard deviation for the sample as

$$S_{IPC} = \frac{S_{IPC}}{\sqrt{N_{cluster}}},\qquad(6.9)$$

The estimated standard error can be used to calculate the *error bounds* and *confidence interval*. Using the properties of the normal distribution, the 95% confidence interval is given by $\mu_{IPC}^{sample} \pm 1.96\, S_{IPC}$, where the error bound is $\pm 1.96 S_{IPC}$. A confidence interval of 95% implies that 95 out of 100 sample estimates may be expected to fit into this interval. Moreover, for a well-designed sample, where non-sampling bias is negligible, the true mean of the population may also be expected to fall within this range. Low standard errors imply little variation in repeated estimates and consequently result in higher precision.

Table 6.3 shows the confidence interval measurements from estimates obtained from single samples ($N_{cluster} = 2000$) with cluster sizes of 3000 instructions. The 95% error bounds are also shown. The ammp benchmark

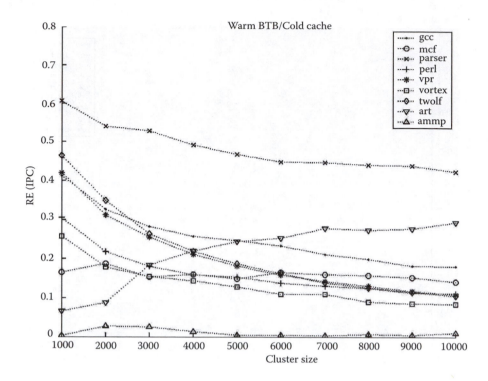

Figure 6.10 Warm BTB/cold cache.

has the maximum standard error and therefore largest error bounds. Its confidence interval indicates that the mean IPC for repeated samples should be between 0.19019 and 0.29761 ($\mu_{IPC}^{sample} \pm CI$). Whether or not the precision provided by this range is acceptable depends on the tolerable error decided upon. The values of the true mean (μ_{IPC}^{true}) are included in Table 6.3 to show that the confidence interval also contains μ_{IPC}^{true}. This is true for all the benchmarks.

Figure 6.12 shows the variability of cluster means across all clusters in the sample. The x-axis represents the cluster number and the y-axis is the mean IPC for each cluster in the sample. This figure provides insights into why some benchmarks are more difficult to sample than others. It shows the distribution of the mean IPCs of the clusters using a 1000-cluster sample. Note that benchmarks with small variations among cluster means, such as vpr, vortex, and twolf, are conducive to accurate sampling. Benchmarks such as gcc, parser, perl, and mcf exhibit high variation in the cluster means and are therefore difficult to sample. It is clear that the precision of a sampling regimen depends upon the homogeneity of the cluster means. For these benchmarks, the number of clusters needs to be large enough to offset the effects of the highly heterogeneous cluster means. However, if variation

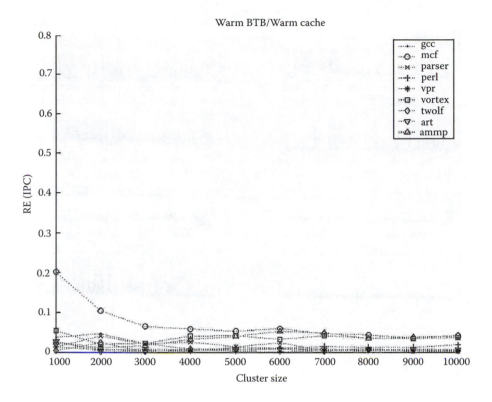

Figure 6.11 Warm BTB/warm cache.

Table 6.3 Sampled estimate accuracy

benchmark	True mean μ_{IPC}^{true}	Estimated mean μ_{IPC}^{sample}	Standard error S_{IPC}	95% Error bound CI	Absolute error $\left\|\mu_{IPC}^{true} - \mu_{IPC}^{sample}\right\|$
gcc	0.87314	0.89178	0.02263	±0.04436	0.01864
mcf	0.20854	0.22202	0.01999	±0.03918	0.01348
parser	1.07389	1.05273	0.01343	±0.02632	0.02116
perl	1.28956	1.28458	0.00761	±0.01493	0.00498
vpr	1.18062	1.17164	0.00601	±0.01178	0.00898
vortex	0.92672	0.92415	0.00487	±0.00955	0.00257
twolf	0.97398	0.97523	0.00599	±0.01175	0.00125
art	0.77980	0.78220	0.01816	±0.03560	0.00240
ammp	0.24811	0.24390	0.02740	±0.05371	0.00421

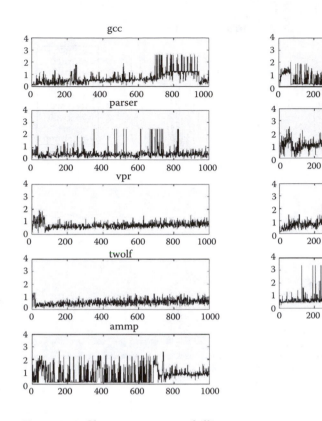

Figure 6.12 Cluster means variabillity.

among the cluster means is too low, then the error bounds will become very small and the μ_{IPC}^{true} may no longer be enveloped by the confidence interval.

The variability of cluster means can also be used to explain some of the strange behavior of the art benchmark. Figure 6.13 shows the variability of the cluster means for the art benchmark for cluster sizes of 1000, 3000, 5000, and 10,000 instructions. As shown in this figure, the smallest cluster size of 1000 exhibits a large degree of variability, including a large number of transient spikes up to an IPC of approximately 3.5. As the cluster sizes increase, the behavior of the cluster means changes. At a 3000-instruction cluster size the number of transient spikes begins to diminish, and at 5000 the spikes only reach an approximate IPC of 2.5. At a 10,000-instruction cluster size the large spikes are all but removed. This suggests a low-pass filtering effect due to larger cluster sizes. For this benchmark, the performance of cluster sampling is more accurate when smaller cluster sizes are used. As the cluster sizes increase, significant transient spikes are removed because the behavior is averaged among a greater number of instructions. Although this suggests that larger cluster sizes can be bad, the positive side is that it helps to hide

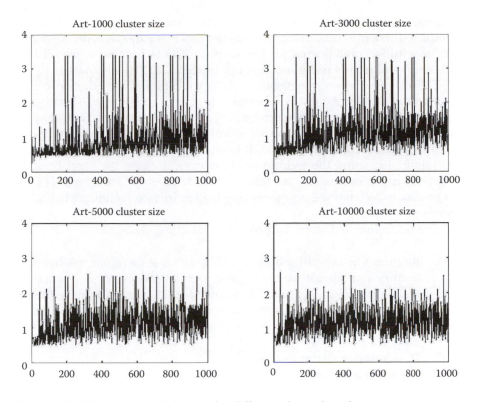

Figure 6.13 Cluster means behavior for different cluster lengths.

the non-sampling bias. The art benchmark performed contrary to expectation by actually increasing in relative error as the cluster sizes increased, but the non-sampling bias was successfully removed in other warm-up policies such as warm BTB–warm cache and stale BTB–warm cache methods.

Because the full-trace simulations are available in this study, it is possible to test whether sample design using standard error achieves accurate results. The estimates of μ_{IPC}^{sample} when compared to μ_{IPC}^{true} show relative errors of less than 2% for most benchmarks (Table 6.3). The conclusion is that a robust sampling regimen can be designed without the need for full-trace simulations. When non-sampling bias is negligible, the sampling regimen can be designed from the data obtained solely from a single sampled run.

6.4   *Concluding remarks*

This chapter has described techniques that have been used in sampling for caches. Although the survey of techniques may not be exhaustive, an attempt has been made to describe some of the more efficient methods in use today. Because techniques for processor simulation have not developed as rapidly, techniques have been developed for accurate processor simulation via systematic reduction in bias. A highly parallel processor model with considerable state information is used for the purpose. The techniques were verified with empirical results using members of the SPEC2000 benchmarks.

The use of the non-sampling bias reduction techniques were demonstrated by sample design for the test benchmarks. To reduce sampling bias, statistical sampling design techniques were employed. The results demonstrate that a regimen for sampling a processor simulation can be developed without the need for full-trace simulations. It is unlikely that all non-sampling bias was eliminated using the techniques. However, because the error bounds calculated using estimated standard error bracketed the true mean IPC, it can be concluded that the non-sampling bias reduction technique is highly effective.

The recommended steps for processor sampling design are

1. **Reduce non-sampling bias:** This requires a state repair mechanism. Empirical evidence from a highly parallel processor with a robust branch predictor suggests selection of 2000 clusters with a cluster size of 3000 or more instructions, with a full warm-up method of the branch predictor and caches while skipping instructions between clusters.
2. **Determine the sample design:**
   a. **Select a number of clusters:** Simulate using a particular number of clusters.
   b. **Determine error bounds:** Estimate standard error (Equations (6.8) and (6.9)) to determine error bounds/precision of the results. If the error is acceptable, the experiments are completed. Otherwise, increase the sample size by increasing the number of clusters, and resimulate until the desired precision is achieved.

The results of this study demonstrate the power of statistical theory adapted for discrete-event simulation.

References

1. Laha, S., Patel, J.A., and Iyer, R.K., Accurate low-cost methods for performance evaluation of cache memory systems, *IEEE Trans. Comput.*, C-37, 1325–1336, Feb. 1988.
2. Stone, H.S., *High-Performance Computer Architecture*, New York, NY: Addison-Wesley, 1990.
3. Wood, D.A., Hill, M.D., and Kessler, R. E., A model for estimating trace-sample miss ratios, in *Proc. ACM SIGMETRICS '91 Conf. on Measurement and Modeling of Comput. Sys.*, 79–89, May 1991.
4. Kessler, R.E., Hill, M.D., and Wood, D.A., A comparison of trace-sampling techniques for multi-megabyte caches, *IEEE Trans. Comput.*, C-43, 664–675, June 1994.

5. Conte, T.M., Systematic computer architecture prototyping, Ph.D. thesis, Department of Electrical and Computer Engineering, University of Illinois, Urbana, Illinois, 1992.

6. Poursepanj, The PowerPC performance modeling methodology, *Communications ACM*, vol. 37, pp. 47–55, June 1994.

7. Lauterbach, G., Accelerating architectural simulation by parallel execution, in *Proc. 27th Hawaii Int'l. Conf. on System Sciences*, (Maui, HI), Jan. 1994.

8. McCall, J.C.H., *Sampling and Statistics Handbook for Research*, Ames, Iowa: Iowa State University Press, 1982.

9. Fu, J.W.C. and Patel, J.H., Trace driven simulation using sampled traces, in *Proc. 27th Hawaii Int'l. Conf. on System Sciences* (Maui, HI), Jan. 1994.

10. Henry, G.T., *Practical Sampling*, Newbury Park, CA: Sage Publications, 1990.

11. Liu, L. and Peir, J., Cache sampling by sets, *IEEE Trans. VLSI Systems*, 1, 98–105, June 1993.

12. McFarling, S., Combining branch predictors, technical report TN-36, Digital Western Research Laboratory, June 1993.

13. Song, S.P. and Denman, M., The PowerPC 604 RISC microprocessor, technical report, Somerset Design Center, Austin, TX, Apr. 1994.

14. Wenish, T.F., Wunderlich, R.E., Falsafi, B., and Hoe, J.C., SMARTS: Accelerating microarchitecture simulation via rigorous statistical sampling, *Proc. 30th ISCA*, 2003.

15. Mangione-Smith, W.H., Abraham, S.G., and Davidson, E.S., Architectural vs Delivered Performance of the IBM RS/6000 and the Astronautics ZS-1, in *Proc. 24th Hawaii International Conference on System Sciences*, January 1991.

16. Lui, L. and Peir, J., Cache sampling by sets, *IEEE Trans. VLSI Systems*, 1, 98–105, June 1993.

17. Conte, T.M., Hirsch, M.A., and Hwu, W.W., Combining trace sampling with single pass methods for efficient cache simulation, *IEEE Transactions on Computers*, C-47, Jun. 1998.

18. Haskins, J.W., and Skadron, K.. Memory reference reuse latency: Accelerated sampled microarchitecture simulation, in *Proc of the 2003 IEEE International Symposium on Performance Analysis of Systems and Software*, 195–203, Mar. 2003.

19. Conte, T.M., Hirsch, M.A., and Menezes, K.N., Reducing state loss for effective trace sampling of superscalar processors, in *Proc of the 1996 International Conference on Computer Design* (Austin, TX), Oct. 1996.

20. Burger, D.C., and Austin, T.M., The simplescalar toolset, version 2.0, *Computer Architecture News*, 25, 3, 13–25, 1997.

Chapter Seven

SimPoint: Picking Representative Samples to Guide Simulation

Brad Calder, Timothy Sherwood, Greg Hamerly and Erez Perelman

Contents

7.1   Introduction

Understanding the cycle-level behavior of a processor during the execution of an application is crucial to modern computer architecture research. To gain this understanding, researchers typically employ detailed simulators that model each and every cycle. Unfortunately, this level of detail comes at the cost of speed, and simulating the full execution of an industry standard benchmark can take weeks or months to complete, even on the fastest of simulators. Exacerbating this problem further is the need of architecture researchers to simulate each benchmark over a variety of different architectural configurations and design options, to find the set of features that provides the appropriate tradeoff between performance, complexity, area, and power. The same program binary, with the exact same input, may be run hundreds or thousands of times to examine how, for example, the effectiveness of a given architecture changes with its cache size. Researchers need techniques that can reduce the number of machine-months required to estimate the impact of an architectural modification without introducing an unacceptable amount of error or excessive simulator complexity.

Executing programs have behaviors that change over time in ways that are not random but rather are often structured as sequences of a small number of reoccurring behaviors that are called *phases*. This structured behavior is a great benefit to simulation. It allows very fast and accurate sampling by identifying each of the repetitive behaviors and then taking only a single sample of each repeating behavior to represent that behavior. All of these representative samples together represent the complete execution of the program. This is the underlying philosophy of the tool called SimPoint [16,17,14,1,9,8]. SimPoint intelligently chooses a very small set of samples called *simulation points* that, when simulated and weighed appropriately, provide an accurate picture of the complete execution of the program. Simulating only these carefully chosen simulation points can save hours of simulation time over statistically random sampling, while still providing the accuracy needed to make reliable decisions based on the outcome of the cycle level simulation. This chapter shows that repetitive phase behavior can be found in programs and describes how SimPoint automatically finds these phases and picks simulation points.

7.2 Defining phase behavior

Because phases are a way of describing the reoccurring behavior of a program executing over time, let us begin the analysis of phases with a demonstration of the time-varying behavior [15] of two different programs from SPEC (Standard Performance Evaluation Cooperative) 2000, gcc and gzip. To characterize the behavior of these programs we have simulated their complete execution from start to finish. Each program executes many billions of instructions, and gathering these results took several machine-months of simulation time. The behavior of each program is shown in Figures 7.1 and 7.4. Each figure shows how the CPI changes for these two programs over time. Each point on the graph represents the average value for CPI taken over a window of 10 million executed instructions (which we call an interval). These graphs show that the average behavior does not sufficiently characterize the behavior of the programs.

Note that not only do the behaviors of the programs change over time, they change on the largest of time scales, and even here we can find repeating behaviors. The programs may have stable behavior for billions of instructions and then change suddenly. In addition to performance, we have found for the SPEC95 and SPEC2000 programs that the behavior of all of the architecture metrics (branch prediction, cache misses, etc.) tend to change in unison, although not necessarily in the same direction [15,17]. This change in unison is due to an underlying change in the program's execution, which can have drastic changes across a variety of architectural metrics. The underlying methodology used in this chapter is the ability to automatically identify these underlying program changes without relying on architectural metrics to group the program's execution into phases. To ground our discussions in a common vocabulary, the following is a list of definitions that

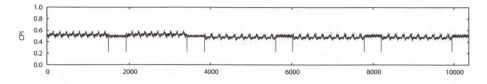

Figure 7.1 Time-varying graphs for CPI from each interval of execution for gz-ip–graphic at 10 million interval size. The x-axis represents the execution of the program over time. The results are nonaccumulative.

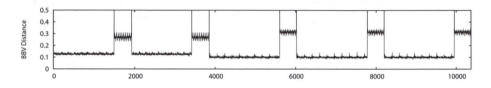

Figure 7.2 Time-varying graph showing the distance to the target vector from each interval of execution in gzip–graphic for an interval size of 10 million instructions. To produce the target vector, we create a basic block vector treating the whole program as one interval. The target vector is a signature of the program's overall behavior.

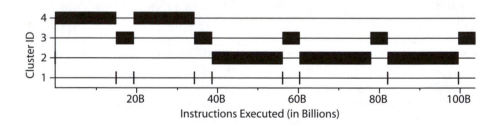

Figure 7.3 Shows which intervals during the program's execution are partitioned into the different phases as determined by the SimPoint phase classification algorithm. The full run of execution is partitioned into a set of four phases.

are used in this chapter to describe program phase behavior and it's automated classification.

- Interval—A section of continuous execution (a slice in time) of a program. For the results in this chapter all intervals are chosen to be the same size, as measured in the number of instructions committed within an interval (e.g., 1, 10, or 100 million instructions [14]). All intervals are assumed to be nonoverlapping, so to perform our analysis we break a program's execution up into contiguous nonoverlapping fixed-length intervals.

- Similarity—Similarity defines how close the behavior of two intervals are to one another as measured across some set of metrics. Well-formed phases should have intervals with similar behavior across various architecture metrics (e.g., IPC, cache misses, branch misprediction).
- Phase—A set of intervals within a program's execution that have all behavior similar to one another, regardless of temporal adjacency. In this way a phase can consist of intervals that re-occur multiple times (repeat) through the execution of the program (as can be seen in gzip and gcc).
- Phase Classification—Phase classification breaks up a program/input's set of intervals into phases with similar behavior. This phase behavior is for a specific program binary running a specific input (a binary/input pair).

7.3 The strong correlation between code and performance

As mentioned in the prior section, for an automated phase analysis technique to be applicable to architecture design space exploration, we must be able to directly identify the underlying changes taking place in the executing program. This section is a description of techniques that have been shown effective at accomplishing this.

7.3.1 Using an architecture-independent metric for phase classification

To find phase information, any effective technique requires a notion of how similar two parts of the execution in a program are to one another. In creating this *similarity* metric it is advantageous not to rely on statistics such as cache miss rates or performance, because this would tie the phases to those statistics. If that was done, then the phases would need to be reanalyzed every time there is a change to some architecture parameter (either statically if the size of the cache changed, or dynamically if some policy is changed adaptively). This is not acceptable, because our goal is to find a set of samples that can be used across an architecture design space exploration. To address this, we need a metric that is *independent* of any particular hardware based statistic, yet it must still relate to the fundamental changes in behavior shown in Figures 7.1 and 7.4.

An effective way to design such a metric is to base it on the behavior of a program in terms of the code that is executed over time. There is a very strong correlation between the set of paths in a program that are executed and the time-varying architectural behavior observed. The intuition behind this is that the code being executed determines the behavior of the program. With this idea it is possible to find the phases in programs using *only* a metric related to how the code is being exercised (i.e., both what code is touched and how often). It is important to understand that this approach can find the same phase behavior shown in Figures 7.1 and 7.4 by examining only the frequency with which the code parts (e.g., basic blocks) are executed over time.

7.3.2   Basic block vector

The *Basic Block Vector* (or BBV) [16] is a structure designed to concisely capture information about how a program is changing behavior over time. A basic block is a section of code that is executed from start to finish with one entry and one exit. The metric for comparing two time intervals in a program is based on the differences in the frequency that each basic block is executed during those two intervals. The intuition behind this is that the behavior of the program at a given time is directly related to the code it is executing during that interval, and basic block distributions provide us with this information.

A program, when run for any interval of time, will execute each basic block a certain number of times. Knowing this information provides a code signature for that interval of execution and shows where the application is spending its time in the code. The general idea is that knowing the basic block distribution for two different intervals gives two separate signatures, which we can then compare to find out how similar the intervals are to one another. If the signatures are similar, then the two intervals spend about the same amount of time in the same code, and the performance of those two intervals should be similar.

More formally, a BBV is a one-dimensional array, with one element in the array for each static basic block in the program. Each interval in an executed program gets one BBV, and at the beginning of each interval we start with a BBV containing all zeros. During each interval, we count the number of times each basic block in the program has been entered (just during that interval), and record that number into the vector (weighed by the number of instructions in the basic block). Therefore, each element in the array is the count of how many times the corresponding basic block has been entered during an interval of execution, multiplied by the number of instructions in that basic block. For example, if the 50th basic block has one instruction and is executed 15 times in an interval, then bbv[50] = 15 for that interval. The BBV is then normalized to 1 by dividing each element by the sum of all the elements in the vector.

We recently examined frequency vector structures other than BBVs for the purpose of phase classification. We have looked at frequency vectors for data, loops, procedures, register usage, instruction mix, and memory behavior [9]. We found that using register usage vectors, which simply counts for a given interval the number of times each register is defined and used, provides similar accuracy to using BBVs. In addition, tracking only loop and procedure branch execution frequencies performed almost as well as using the full basic block information. We also found, for SPEC2000 programs, that creating data vectors or combined code and data vectors did not improve classification over just using code [9].

7.3.3   Basic block vector difference

In order to find patterns in the program we must first have some way of comparing the similarity of two BBVs. The operation needed takes as input two BBVs and outputs a single number corresponding to how similar they are.

We use BBVs to compare the intervals of the application's execution. The intuition behind this is that the behavior of the program at a given time is directly related to the code executed during that interval [16]. We use the BBVs as signatures for each interval of execution: each vector tells us what portions of code are executed, and how frequently those portions of code are executed. By comparing the BBVs of two intervals, we can evaluate the similarity of the two intervals. If two intervals have similar BBVs, then the two intervals spend about the same amount of time in roughly the same code, and therefore we expect the performance of those two intervals to be similar.

There are several ways of comparing two vectors to one another, such as taking the dot product or finding the Euclidean or Manhattan distance.

The Euclidean distance, which has been shown to be effective for offline phase analysis [17,14], can be found by treating each vector as a single point in a D-dimensional space, and finding the straight-line distance between the two points. More formally, the Euclidean distance of two vectors a and b in D-dimensional space is given by

$$EuclideanDist(a,b) = \sqrt{\sum_{i=1}^{D} (a_i - b_i)^2}$$

The Manhattan distance, on the other hand, is the distance between two points if the only paths followed are parallel to the axes, and is more efficient for on-the-fly phase analysis [18,10]. In two dimensions, this is analogous to the distance traveled by a car in a city through a grid of city streets. This has the advantage that it always gives equal weight to each dimension. The Manhattan distance is computed by summing the absolute value of the element-wise subtraction of two vectors. For vectors a and b in D-dimensional space, the distance is

$$ManhattanDist(a,b) = \sum_{i=1}^{D} |a_i - b_i|$$

7.3.4 Showing the correlation between code signatures and performance

A detailed study showing that there is a strong correlation between code and performance can be found in Lau et al. [8]. The graphs in Figures 7.4 and 7.5 give one representation of this by showing the time- varying CPI and BBV distance graphs for gcc-166 right next to each other. The time-varying CPI graph plots the CPI for each interval executed (at 10M interval size) showing how the program's CPI varies over time. Similarly, the BBV distance graph plots for each interval the Manhattan distance of the BBV (code signature) for that interval from the whole program target vector. The whole program target vector is the BBV if the whole program is viewed as a single interval. The same information is also provided for gzip in Figures 7.1 and 7.2. The time-varying graphs show that changes in CPI have corresponding changes in code signatures, which is one indication of strong phase behavior for these applications. These results show that the BBV can accurately track the changes in CPI for both gcc and gzip. It is easy to see that over time the CPI changes accurately mirror changes visible in the BBV distance graph.

These plots show that code signatures have a strong correlation to the changes in CPI even for complex programs such as gcc. The results for gzip show that the phase behavior can be found even if the intervals' CPIs have small variance. This brings up an important point about picking samples for simulation based on code vectors versus CPI or some other hardware metric. Assume we have two intervals with *different code signatures* but they have very *similar CPIs* because both of their working sets fit completely in the cache. During a design space exploration search, as the cache size changes, the two interval CPIs may differ drastically because one of them no longer fits into the cache. This is why it is important to perform the phase analysis by comparing the code signatures independent of the underlying architecture, and not based upon CPI thresholds. We have found that the BBV code signatures correctly identify this difference, which cannot be seen by looking at just the CPI. If the purpose of a study is to perform design space exploration, it is important to be able to pick samples that will be representative of the program's execution no matter the underlying architecture configuration. See Lau, Sampson, et al. [8], for a complete discussion and analysis on the strong correlation between code and performance.

7.4 Automatically finding phase behavior

Frequency vectors (BBVs, vectors based on the execution of loops and procedures, or some other behavior discussed in Lau, Schoenmackers, and Calder [9]) provide a compact and representative summary of a program's behavior for each interval of execution. By examining the similarity between them, it is clear that there are high-level patterns in each program's execution. In this section we describe the algorithms used to automatically detect these patterns.

7.4.1 Using clustering for phase classification

It is extremely useful to have an automated way of extracting phase information from programs. Clustering algorithms from the field of machine learning have been shown to be very effective [17] at breaking the complete execution of a program into phases that have similar frequency vectors. Because the frequency vectors correlate to the overall performance of the program, grouping intervals based on their frequency vectors produces phases that are similar not only in the distribution of program structures used, but also in every other architecture metric measured, including overall performance.

The goal of clustering is to divide a set of points into groups, or clusters, such that points within each cluster are similar to one another (by some metric, usually distance), and points in different clusters are different from one another. The k-means algorithm [11] is an efficient and well-known clustering algorithm, which we use to quickly and accurately split program behavior into phases. We use random linear projection [5] to reduce the dimension of the input vectors while preserving the underlying similarity information, which speeds up the execution of k-means. One drawback of the k-means algorithm is that it requires the number of clusters k as an input to the algorithm, but we do not know beforehand what value is appropriate. To address this, we run the algorithm for several values of k, and then use a goodness score to guide our final choice for k.

Taking this to the extreme, if every interval of execution is given its very own cluster, then every cluster will have perfect homogeneous behavior. Our goal is to choose a clustering with a *minimum number of clusters* where each cluster has a certain level of homogeneous behavior.

The following steps summarize the phase clustering algorithm at a high level. We refer the interested reader to Sherwood et al. [17] for a more detailed description of each step.

1. Profile the program by dividing the program's execution into contiguous intervals of size N (e.g., 1 million, 10 million, or 100 million instructions). For each interval, collect a frequency vector tracking the program's use of some program structure (basic blocks, loops, register usage, etc.). This generates a frequency vector for every interval. Each frequency vector is normalized so that the sum of all the elements equals 1.

2. Reduce the dimensionality of the frequency vector data to D dimensions using random linear projection. The advantage of performing clustering on projected data is that it speeds up the k-means algorithm significantly and reduces the memory requirements by several orders of magnitude over using the original vectors, while preserving the essential similarity information.

3. Run the *k*-means clustering algorithm on the reduced dimensional data with values of *k* from 1 to *K*, where *K* is the maximum number of phases that can be detected. Each run of *k*-means produces a clustering, which is a partition of the data into *k* different phases/clusters. Each run of *k*-means begins with a random initialization step, which requires a random seed.

4. To compare and evaluate the different clusters formed for different *k*, we use the *Bayesian Information Criterion* (BIC) [13] as a measure of the goodness of fit of a clustering to a dataset. More formally, the BIC is an approximation to the probability of the clustering given the data that has been clustered. Thus, the higher the BIC score, the higher the probability that the clustering is a good fit to the data. For each clustering ($k = 1 \ldots K$), the fitness of the clustering is scored using the BIC formulation given in Pelleg and Moore [13].

5. The final step is to choose the clustering with the smallest *k*, such that its BIC score is at least *B*% as good as the best score. The clustering *k* chosen is the final grouping of intervals into phases.

The preceding algorithm groups intervals into phases. We use the Euclidean distance between vectors as our similarity metric. This algorithm has several important parameters (*N*, *D*, *K*, *B*, and more), which must be tuned to create accurate and representative simulation points using SimPoint. We discuss these parameters in more detail later in this chapter.

7.4.2   *Clusters and phase behavior*

Figures 7.3 and 7.6 show the result of running the clustering algorithm on gzip and gcc using an interval size of 100 million and setting the maximum number of phases (*K*) to 10. The *x*-axis corresponds to the execution of the program in billions of instructions, and each interval is tagged to be in one of the clusters (labeled on the *y*-axis).

For gzip, the full run of the execution is partitioned into a set of four clusters. Looking at Figure 7.2 for comparison, the cluster behavior captured by the offline algorithm lines up quite closely with the behavior of the program. Clusters 2 and 4 represent the large sections of execution that are similar to one another. Cluster 3 captures the smaller phase that lies in between these larger phases. Cluster 1 represents the phase transitions between the three dominant phases. The cluster 1 intervals are grouped into the same phase because they execute a similar combination of code, which happens to be part of code behavior in either cluster 2 or 4 and part of code executed in cluster 3. These transition points in cluster 1 also correspond to the same intervals that have large cache miss rate spikes seen in the time-varying graphs of Figure 7.1.

Figure 7.6 shows how gcc is partitioned into eight different clusters. In comparing Figure 7.6 to Figures 7.4 and 7.5, we see that even the more complicated behavior of gcc is captured correctly by SimPoint. The dominant

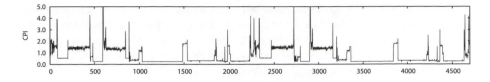

Figure 7.4 Time-varying graphs for CPI from each interval of execution for gcc-166 at 10 million interval size. The x-axis represents the execution of the program over time. The results are nonaccumulative.

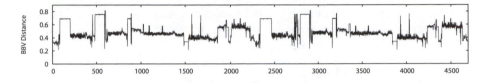

Figure 7.5 Time-varying graph showing the distance to the target vector from each interval of execution in gcc-166 for an interval size of 10 million instructions. To produce the target vector, we create a basic block vector treating the whole program as one interval. The target vector is a signature of the program's overall behavior.

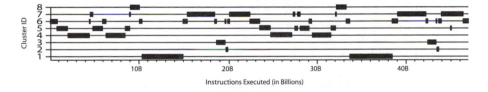

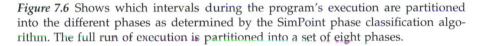

Instructions Executed (in Billions)

Figure 7.6 Shows which intervals during the program's execution are partitioned into the different phases as determined by the SimPoint phase classification algorithm. The full run of execution is partitioned into a set of eight phases.

behaviors in the time-varying CPI and vector distance graphs can be seen grouped together in the dominant phases 1, 4, and 7.

7.5 Choosing simulation points from the phase classification

After the phase classification algorithm described in the previous section has done its job, intervals with similar code usage will be grouped together into the same phase, or cluster. Then from each phase, we choose one representative interval that will be simulated in detail to represent the behavior of the whole phase. Therefore, by simulating *only* one representative interval per phase, we can extrapolate and capture the behavior of the entire program.

To choose a representative, SimPoint picks the interval that is closest to the center of each cluster. The center is the average of all the intervals in the cluster, and is called the *centroid*. This is analogous to the balance point of all the points that are in that cluster, if all points had the same mass. It can also be viewed as the interval that behaves most like the average behavior of the entire phase. Most likely there is no interval that exactly matches the centroid, so the interval closest to the center is chosen. The selected interval is called a *simulation point* for that phase [14,17]. Detailed simulation is then performed on the set of simulation points.

SimPoint also gives a weight for each simulation point. Each weight is a fraction; it is the total number of instructions counting all of the intervals in the cluster, from which the simulation point was taken, divided by the number of instructions in the program. With the weights and the detailed simulation results of each simulation point, we compute a weighted average for the architecture metric of interest (CPI, miss rate, etc). This weighted average of the simulation points gives an accurate representation of the complete execution of the program/input pair.

7.6   Using the simulation points

After the SimPoint algorithm has chosen a set of simulation points and their respective weights, they can be used to accurately estimate the full execution of a program. The next step is to simulate in detail the interval for each simulation point, to collect the desired performance statistics.

7.6.1   Simulation point representation

SimPoint provides the simulation points in two forms:

- **Simulation Point Interval Number**—The interval number for each simulation point is given. The interval numbers are relative to the start of execution, not to the previous simulation point. To get the start of a simulation point, subtract 1 from the interval number, and multiply by the interval size. For example, interval number 15 with an interval size of 10 million instruction means that the simulation point starts at instruction 140 million (i.e., (15 − 1)∗10M) from the start of execution. Detailed simulation of this simulation point would occur from instruction 140 million until just before 150 million.
- **Start PC with Execution Count**—SimPoint also provides for each simulation point the program counter for the first instruction executed for the interval and the number of times that instruction needs to be executed before starting simulation. For example, if the PC is 0×12000340 with an execution count of 1000, then detailed simulation starts the 1000th time that PC is seen during execution, and simulation occurs for the length of the profile interval.

It is highly recommended that you use the simulation point PCs for performing your simulations. There are two reasons for this. The first reason deals with making sure you calculate the instructions during fast-forwarding exactly the same as when the simulation points were gathered. The second reason is that there can be slight variations in execution count between different runs of the same binary/input due to environment variables or operating system variations when running on a cluster of machines. Both of these are discussed in more detail later in this chapter.

7.6.2   Getting to the starting sample image

After choosing the form of simulation points to use, each simulation point is then simulated. Two standard approaches for doing this are to use either fast-forwarding or checkpointing.

7.6.2.1   Fast-forwarding

Sort the simulation points in chronological order. Fast-forward to the start of the first simulation point. Simulate at the desired detail for the size of the interval. Repeat these steps, fast-forwarding from one point to the next combined with detailed simulation, until all simulation intervals have been collected.

7.6.2.2   Checkpointing starting sample image

One advantage of SimPoint is that the state of a program can be checkpointed (e.g., using SimpleScalar's checkpoint facility) right before the start of each simulation point. This checkpointing allows parallel simulation of all of the simulation points at once.

7.6.2.3   Reduced checkpoints

Checkpointing is used to obtain the startup image size of the sample to be simulated. A technique proposed by Van Biesbrouck et al. [1] examines only storing the memory words accessed in the simulation point to create a reduced checkpoint. This results in two orders of magnitude less storage then full checkpointing, and significantly faster simulation.

7.6.3   Warm-up

Using small interval sizes for your simulation points requires having an approach for warming up the architecture state (e.g., the caches, TLBs, and branch predictor). The following are some standard approaches for dealing with warm-up.

7.6.3.1   No warm-up

If a large enough interval size is used (e.g., larger than 100 million instructions), no warm-up may be necessary for many programs. This is the approach used by Intel's PinPoint for simulation [12]. They simulate intervals of size 250 million instructions so they do not have to worry about any warm-up issues. They chose to go the SimPoint route with large interval sizes because of the complexity of integrating statistical simulation and warm-up into their detailed cycle accurate simulator.

7.6.3.2   Assume hit (remove cold structure misses)

All of the large architecture structures (e.g., cache, branch predictors) make use of a warm-up bit that indicates when the first time an entry (e.g., cache block) in that structure is used. If it is the first time, the access is assumed to be a hit or a correct prediction, because most programs have low miss rates. One can also use a miss rate percentage (e.g., 10%) for these cold structure misses, randomly assuming some percentage of the cold start accesses are misses. This a very simple method that provides fairly accurate warm-up state, because the miss rates for these structures are usually fairly low [19,7].

7.6.3.3   Stale state

This is a method of not resetting the architecture structures between simulation points, and instead they are used in the state they were in at the end of the prior simulation point we just fast-forwarded from [4].

7.6.3.4   Calculated warm-up

One can calculate the working set of the most recently accessed data, code, and branch addresses before a simulation point. Then start the simulation of architectural components W instructions before the simulation point, where W is large enough to capture the working set size held by the architecture structures. After these W instructions are simulated, all statistics are reset and detailed simulation starts at that point. The goal of this approach is to bring the working set back into the architecture structures before starting the detailed simulation [3,6].

7.6.3.5   Continuously warm

This approach continuously keeps the state of certain architecture components (e.g., caches) warm even during fast-forwarding [20]. This is feasible if an infrastructure provides fast functional and structure simulation during fast-forwarding. Keeping the cache structures warm will increase the time it takes to perform fast-forwarding, but it is very accurate.

7.6.3.6   Architecture structure checkpoint

An architecture checkpoint is the checkpoint of the potential contents of the major architecture components (caches, branch predictors, etc) at the start of the simulation point [1]. This can be used to significantly reduce warm-up time, because warm-up consists of just reading the architecture structure checkpoint from the file and using it to initialize the architecture structures.

   If you decide to use small interval sizes, *calculated warm-up* and *architecture checkpointing* provide the most accurate and efficient warm-up, although we have found that for many programs *assume hit* and *stale state* are fairly accurate.

7.6.4   Combining the simulation point results

The final step in using SimPoint is to combine the weighted simulation points to arrive at an overall performance estimate for the program's execution.

One cannot just use the standard mean for computing the overall miss rate, because we need to apply a weight to each sample.

Each weight represents the proportion of the total execution that belongs to its phase. The overall performance estimate is the weighted average of the set of simulation point estimates. For example, if we have three simulation points and their weights are [22, 33, 45] and their CPIs are (CPI1, CPI2, CPI3), then the weighted average of these points is

$$CPI = 0.22*CPI1 + 0.33*CPI2 + 0.45*CPI3$$

The weighted average CPI is the estimate of the CPI for the full execution.

7.6.5 Pitfalls to watch for when using simulation points

There are a few important potential pitfalls worth addressing to ensure accurate use of SimPoint's simulation points.

7.6.5.1 Calculating weighted IPC

For IPC (instructions/cycle) we cannot just apply the weights as in the preceding example. We first would need to convert all the simulated samples to CPI before computing the weighted average as given earlier, and then convert the result back to IPC.

7.6.5.2 Calculating weighted miss rates

To compute an overall miss rate, first we must calculate both the weighted average of the number of cache accesses, and the weighted average of the number of cache misses. Dividing the second number by the first gives the cache miss rate. In general, care must be taken when dealing with any ratio because both the numerator and the denominator must be averaged separately and *then* divided.

7.6.5.3 Accurate instruction counts (no-ops)

It is important to count instructions exactly the same for the BBV profiles as for the detailed simulation, otherwise they will diverge. Note that the simulation points on the SimPoint Web site include only correct path instructions and the instruction counts include no-ops. Therefore, to reach a simulation point in a simulator, *every* committed instruction (including no-ops) must be counted.

7.6.5.4 System call effects

Some users have reported system call effects when running the same simulation points under slightly different OS configurations on a cluster of machines. This can result is slightly more or less instructions being executed to get to the same point in the program's execution, and if the number of instructions executed is used to find the simulation point this may lead to variations in the results. To avoid this, we suggest using the Start PC and Execution Count for each simulation point as described above. Another way to avoid variations in startup is to use checkpointing as described above.

7.6.6   Accuracy of simpoint

We now show the accuracy of using SimPoint for the complete SPEC2000 benchmark suite and their reference inputs. Figure 7.7 shows the simulation accuracy results using SimPoint for the SPEC2000 programs when compared to the complete execution of the programs. For these results we use an interval size of 100 million and limit the maximum number of simulation points (clusters) to no more than 10 for the offline algorithm. With the given parameters SimPoint finds four phases for gzip, and eight for gcc. As described earlier, one simulation point is chosen for each cluster, so this means that a total of 400 million instructions were simulated for gzip. The results show that this results in only a 4% error in performance estimation for gzip. Note, if you desire lower error rates, you should use smaller interval sizes and more clusters as shown in Perelman, Hamerly, and Calder [14].

For the non-SimPoint results, we ran a simulation for the same number of instructions as the SimPoint data to provide a fair comparison. The results in Figure 7.7 show that starting simulation at the start of the program results in a median error of 58% when compared to the full simulation of the program, whereas blindly fast forwarding for 1 billion instructions results in a median 23% IPC error. When using the clustering algorithm to create multiple simulation points, we saw a median IPC error of 2%, and an average IPC error of 3%. In comparison to random sampling approaches, we have found that SimPoint is able to achieve similar error rates requiring significantly (five times) less simulation (fast-forwarding) time [14]. In addition, statistical sampling can be combined with SimPoint to create a phase clustering that has a

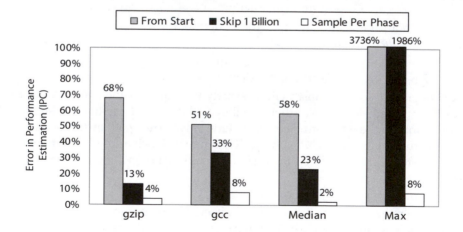

Figure 7.7 Simulation accuracy for the SPEC2000 benchmark suite when performing detailed simulation for several hundred million instructions compared to simulating the entire execution of the program. Results are shown for simulating from the start of the program's execution, for fast-forwarding 1 billion instructions before simulating, and for using SimPoint to choose less than 10 100-million intervals to simulate. The median results are for the complete SPEC2000 benchmarks.

low per-phase variance [14]. Recently, using phase information has even been applied to create accurate and efficient simulation for multi-program work-loads for simultaneous multithreading [2].

7.6.7 Relative error during design space exploration

The absolute error of a program/input run on one hardware configuration is not as important as tracking the change in metrics across different architecture configurations. There is a lot of discussion and research into getting lower error rates. But what often is not discussed is that a low error rate for a single configuration is not as important as achieving the same relative error rates across the design space search and having them all biased in the same direction.

We now examine how SimPoint tracks the relative change in hardware metrics across several different architecture configurations. To examine the independence of the simulation points from the underlying architecture, we used the simulation points for the SimPoint algorithm with a 1 million interval size and max K set to 300. For the program/input runs we examine, we performed full program simulations while varying the memory hierarchy, and for every run we used the same set of simulation points when calculating the SimPoint estimates. We varied the configurations and the latencies of the L1 and L2 caches as described in Perelman, Hamerly, and Calder [14].

Figure 7.8 shows the results across the 19 different architecture configu-rations for gcc-166. The left y-axis represents the performance in instructions

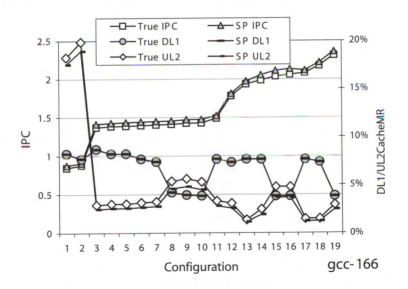

Figure 7.8 This plot shows the true and estimated IPC and cache miss rates for 19 different architecture configurations for the program gcc. The left y-axis is for the IPC and the right y-axis is for the cache miss rates for the L1 data cache and unified L2 cache. Results are shown for the complete execution of the configuration and when using SimPoint.

per cycle and the *x*-axis represents different memory configurations from the baseline architecture. The right *y*-axis shows the miss rates for the data cache and unified L2 cache, and the L2 miss rate is a local miss rate. For each metric, two lines are shown, one for the true metric from the *complete* detailed simulation for every configuration, and the second for the estimated metric using our simulation points. For each graph, the configurations on the *x*-axis are sorted by the IPC of the full run.

Figure 7.8 shows that the simulation points, which are chosen by only looking at code usage, can be used across different architecture configurations to make accurate architecture design trade-off decisions and comparisons. These results show that simulation points track the relative changes in performance metrics between configurations. One interesting observation is that although the simulation results from SimPoint have a bias in the metrics, this bias is consistent and always in the same direction across the different configurations for a given program/input run. This is true for both IPC and cache miss rates. One reason for this bias is that SimPoint chooses the most representative interval from each phase, and intervals that represent phase change boundaries may (if they occur enough) or may not (if they do not occur enough) be represented by a simulation point.

7.7 Discussion about running simpoint

The SimPoint toolkit implements the algorithms described in this chapter. There are a variety of parameters that can be tuned when running the tool to create simulation points for new benchmarks, architectures, or inputs. In this section, we describe these parameters and discuss how they may be adjusted to meet your simulation needs.

7.7.1 Size of interval

The number of instructions per interval is the granularity of the algorithm. The interval size directly relates to the number of intervals, because the dynamic program length is the number of intervals times the interval size. Larger intervals allow more aggregate profile (basic block vector) representations of the program, whereas smaller intervals allow for more fine-grained representations. The interval size affects the number of simulation points; with smaller intervals more simulation points are needed than when using larger intervals to represent the same proportion of the program. Perelman et al. [14] showed that using smaller interval sizes (1 million or 10 million) results in more accuracy when using SimPoint and less simulation time. The disadvantage is that with smaller interval sizes warm-up becomes more of an issue, whereas with larger interval sizes warm-up is not as much of an issue and may be preferred for some simulation environments [12].

7.7.2 Number of intervals

There should be a fair number of intervals for the clustering algorithm to choose from. A good rule of thumb is to make sure to use at least 1000 intervals in order for the clustering algorithm to be able to find a good partition of the intervals. If there are too few intervals, then decrease the interval size to obtain more intervals for clustering.

7.7.3 Number of clusters (K)

The maximum number of clusters (K), along with the interval size, represents the maximum amount of simulation time that will be needed when looking to choose simulation points. If SimPoint chooses a number of clusters that is close to the maximum allowed, then it is possible that K is too small. If this is the case and more simulation time is acceptable, it is better to double the K and rerun the SimPoint analysis.

Creating simulation points with SimPoint comes down to recognizing the tradeoff of accuracy for simulation time. If a user wants to place a low limit on the number of clusters to limit simulation time, SimPoint can still provide accurate results, but some intervals with differing behaviors may be clustered together as a result.

7.7.4 Random seeds

The k-means clustering algorithm starts from a randomized initialization, which requires a random seed. It is well-known that k-means can produce very different results depending on its initialization, so it is good to use many different random seeds for initializing different k-means clusterings, and then allow SimPoint to choose the best clustering. We have found that in practice, using five to seven random seeds works well.

7.7.5 Number of iterations

The k-means algorithm iterates either until it hits a maximum number of iterations or until it reaches a point where no further improvement is possible (whichever is less). In most cases 100 iterations is sufficient for the maximum number, but more may be required, especially if the number of intervals is very large compared to the number of clusters. A very rough rule of thumb is the number of iterations should be set to $\sqrt{N/k}$, where N is the number of intervals and k is the number of clusters.

7.7.6 Number of dimensions

SimPoint uses random linear projection to reduce the dimension of the clustered data, which dramatically reduces computational requirements while retaining the essential similarity information. SimPoint allows the user to define the number of dimensions to project down to. In our experiments we project down to 15 dimensions, as we have found that using it produces the same phases as using the full dimension. We believe this to be adequate for SPEC2000 applications, but it is possible to test other values by looking at the consistency of the clusters produced when using different dimensions [17].

7.7.7 BIC percentage

The BIC gives a measure of the goodness of the clustering of a set of data, and BIC scores can be compared for different clusterings of the same data. However, the BIC score is an approximation of a probability, and often increases as the number of clusters increase. This can lead to often selecting the clustering with the most clusters. Therefore, we look at the range of BIC scores and select the score that attains some high percentage of this range (e.g., we use 90%). When the BIC rises and then levels off, this method chooses a clustering with the fewest clusters that is near the maximum value. Choosing a lower BIC percentage would prefer fewer clusters, but at the risk of less accurate simulation.

7.8 Summary

Understanding the cycle level behavior of a processor running an application is crucial to modern computer architecture research, and gaining this under-standing can be done efficiently by judiciously applying detailed cycle level simulation to only a few simulation points. The level of detail provided by cycle-level simulation comes at the cost of simulation speed, but by targeting only one or a few carefully chosen samples for each of the small number of behaviors found in real programs, this cost can be reduced to a reasonable level.

The main idea behind SimPoint is the realization that programs typically only exhibit a few unique behaviors that are interleaved with one another through time. By finding these behaviors and then determining the relative importance of each one, we can maintain a high-level picture of the pro-gram's execution and at the same time quantify the cycle-level interaction between the application and the architecture. The key to being able to find these phases in an efficient and robust manner is the development of a metric that can capture the underlying shifts in a program's execution that result in the changes in observed behavior. In this chapter we have discussed one such method of quantifying executed code similarity and use it to find program phases through the application of statistical and machine learning methods.

The methods described in this chapter are distributed as part of SimPoint [14,17]. SimPoint automates the process of picking simulation points using an offline phase classification algorithm, which significantly reduces the amount of simulation time required. These goals are met by simulating only a handful of *intelligently* picked sections of the full program. When these simulation points are carefully chosen, they provide an accurate picture of the complete execution of a program, which gives a highly accurate estima-tion of performance. The SimPoint software can be downloaded at http://www.cse.ucsd.edu/users/calder/simpoint/.

Acknowledgments

This work was funded in part by NSF grant No. CCR-0311710, NSF grant No. ACR-0342522, a UC MICRO grant, and a grant from Intel and Microsoft.

References

1. Van Biesbrouck, M., Eeckhout, L., and Calder, B., Efficient sampling startup for uniprocessor and simultaneous multithreading simulation. Technical Report UCSD-CS2004-0803, University of California at San Diego, November 2004.
2. Van Biesbrouck, M., Sherwood, T., and Calder, B., A co-phase matrix to guide simultaneous multithreading simulation, in *IEEE International Symposium on Performance Analysis of Systems and Software*, March 2004.
3. Conte, T.M., Hirsch, M.A., and Hwu, W.W., Combining trace sampling with single pass methods for efficient cache simulation, *IEEE Transactions on Computers* 47, 6, 714–720, 1998.
4. Conte, T.M., Hirsch, M.A., and Menezes, K.N., Reducing state loss for effective trace sampling of superscalar processors, in *Proceedings of the 1996 International Conference on Computer Design (ICCD)*, October 1996.
5. Dasgupta, S., Experiments with random projection. In *Uncertainty in Artificial Intelligence: Proceedings of the Sixteenth Conference* (UAI-2000), 143–151, 2000.
6. Haskins, J. and Skadron, K., Memory reference reuse latency: Accelerated sampled microarchitecture simulation, in *Proceedings of the 2003 IEEE International Symposium on Performance Analysis of Systems and Software*, March 2003.
7. Kessler, R.E., Hill, M.D., and Wood, D.A., A comparison of trace-sampling techniques for multi-megabyte caches, *IEEE Transactions on Computers* 43, 6, 664–675, 1994.
8. Lau, J., Sampson, J., Perelman, E., Hamerly, G., and Calder, B., The strong correlation between code signatures and performance, in *IEEE International Symposium on Performance Analysis of Systems and Software*, March 2005.
9. Lau, J., Schoenmackers, S., and Calder, B., Structures for phase classification, in *IEEE International Symposium on Performance Analysis of Systems and Software*, March 2004.
10. Lau, J., Schoenmackers, S., and Calder, B., Transition phase classification and prediction, in *The International Symposium on High Performance Computer Architecture*, February 2005.
11. MacQueen, J., Some methods for classification and analysis of multivariate observations, in L.M. LeCam and J.Neyman, editors, *Proceedings of the Fifth Berkeley Symposium on Mathematical Statistics and Probability*, volume 1, 281–297, University of California Press: Berkeley, 1967.
12. Patil, H., Cohn, R., Charney, M., Kapoor, R., Sun, A., and Karunanidhi, A., Pinpointing representative portions of large Intel Itanium programs with dynamic instrumentation, in *International Symposium on Microarchitecture*, December 2004.
13. Pelleg, D. and Moore, A., X-means: Extending K-means with efficient estimation of the number of clusters, in *Proceedings of the 17th International Conf. on Machine Learning*, 727–734, 2000.

14. Perelman, E., Hamerly, G., and Calder, B., Picking statistically valid and early simulation points, in *International Conference on Parallel Architectures and Compilation Techniques*, September 2003.
15. Sherwood, T. and Calder, B., Time varying behavior of programs, Technical Report UCSD-CS99-630, University of California at San Diego, August 1999.
16. Sherwood, T., Perelman, E., and Calder, B., Basic block distribution analysis to find periodic behavior and simulation points in applications, in *International Conference on Parallel Architectures and Compilation Techniques*, September 2001.
17. Sherwood, T., Perelman, E., Hamerly, G., and Calder, B., Automatically characterizing large scale program behavior, in *The International Conference on Architectural Support for Programming*, October 2002.
18. Sherwood, T., Sair, S., and Calder, B., Phase tracking and prediction, in *The Annual International Symposium on Computer Architecture*, June 2003.
19. Wood, D.A., Hill, M.D., and Kessler, R.E., A model for estimating trace-sample miss ratios, in *ACM SIGMETRICS International Conference on Measurement and Modeling of Computer Systems*, May 1991.
20. Wunderlich, R., Wenisch, T., Falsafi, B., and Hoe, J., Smarts: Accelerating microarchitecture simulation via rigorous statistical sampling, in *The Annual International Symposium on Computer Architecture*, June 2003.

Chapter Eight

Statistical Simulation

Lieven Eeckhout

Contents

8.1 Introduction

Computer system design is an extremely time-consuming, complex process, and simulation has become an essential part of the overall design activity. Simulation is performed at many levels, from circuits to systems, and at different degrees of detail as the design evolves. Consequently, the designer's toolbox holds a number of evaluation tools, often used in combination, that have different complexity, accuracy, and execution time properties.

For simulation at the microarchitecture level, detailed models of register transfer activity are typically employed. These simulators track instructions and data on a clock-cycle basis and typically provide detailed models for features such as instruction issue mechanisms, caches, load/store queues, and branch predictors, as well as their interactions. For input, microarchitecture simulators take sets of benchmark programs including both standard and company proprietary suites. These benchmarks may each contain billions of dynamically executed instructions, and typical simulators run many orders of magnitude slower than real processors. The result is a relatively long runtime for even a single simulation.

However, processor simulation at such a high level of detail is not always appropriate, nor is it called for. For example, early in the design process, when the design space is being explored and a high-level microarchitecture is being determined, too much detail is unnecessary. When a processor microarchitecture is initially being defined, a number of basic design decisions need to be made. These decisions involve basic cycle time and instruction per cycle (IPC) tradeoffs, cache and predictor sizing tradeoffs, and performance/power tradeoffs. At this stage of the design process, detailed microarchitecture simulations of specific benchmarks aren't feasible for a number of reasons. For one, the detailed simulator itself takes considerable time and effort to develop. Second, benchmarks restrict the studied application space being evaluated to those specific programs. To study a fairly broad design space, the number of simulation runs can be quite large. Finally, highly accurate performance estimates are illusory, anyway, given the level of design detail that is actually known.

Similarly, for making system-level design decisions, where a processor (or several processors) may be combined with many other components, a very detailed simulation model is often unjustified and/or impractical. Even though the detailed processor microarchitecture may be known, the simulation complexity is multiplied many fold by the number of processors and by larger benchmark programs typically required for studying system-level behavior.

This chapter describes statistical simulation that can be used to overcome many of the shortcomings of detailed simulation for those situations where detailed modeling is impractical, or at least overly time consuming. Statistical simulation measures a well-chosen set of characteristics during program execution, generates a synthetic trace with those characteristics, and simulates the synthetic trace. If the set of characteristics reflects the key properties of the program's behavior, accurate performance/power predictions can be made. The statistically generated synthetic trace is several orders of magnitude smaller than the original program execution, hence simulation finishes very quickly. The goal of statistical simulation is not to replace detailed simulation but to be a useful complement. Statistical simulation can be used to identify a region of interest in a large design space that can, in turn, be further analyzed through slower but more detailed architectural simulations. In addition, statistical simulation requires relatively little new tool development effort. Finally, it provides a simple way of modeling superscalar processors as components in large-scale systems where very high detail is not required or practical.

This chapter is organized as follows. It first describes statistical simulation and provides an evaluation of its speed and accuracy in Section 8.2. Section 8.3 discusses a number of applications for statistical simulation. Previous work done on statistical simulation is discussed in Section 8.4. This chapter is summarized in Section 8.5.

8.2   Statistical simulation

Statistical simulation [2,6,8,9,10,17,18,19,20] consists of three steps, as is shown in Figure 8.1. In the first step, a collection of program execution characteristics is measured—this is done through specialized cache and predictor simulation, which we call *statistical profiling*. Subsequently, the obtained *statistical profile* is used to generate a *synthetic trace*. In the final step, this synthetic trace is simulated on a trace-driven simulator. The following subsections outline all three steps.

8.2.1   Statistical profiling

In a statistical profile, there is a distinction between microarchitecture-independent characteristics and microarchitecture-dependent characteristics.

8.2.1.1   Microarchitecture-independent characteristics

During statistical profiling we build a *statistical flow graph (SFG)*. To clarify how this is done, see Figure 8.2 in which a first-order SFG is shown for an example basic block sequence AABAABCABC. Each node in the graph represents the current basic block. This is shown through the labels A, B, and C in the first-order SFG. The value in each node shows the *occurrence* or the number of times the node appears in the basic block stream. For example, basic block A appears five times in the example basic block sequence; consequently, the occurrence for node A equals 5. The percentages next to the edges represent the

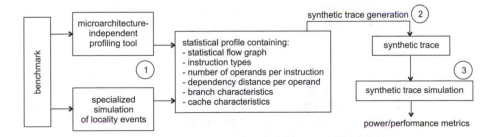

Figure 8.1 Statistical simulation: framework. Reprinted with permission from [8] © 2004 IEEE.

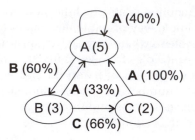

Figure 8.2 Example first-order statistical flow graph (SFG) for basic block sequence AABAABCABC.

transition probabilities $\Pr[B_n \mid B_{n-1}]$ between the nodes. For example, there are 40% and 60% probabilities to execute basic block A and B, respectively, after executing basic block A. Eeckhout et al. [8] studied higher-order SFGs and found that a first-order SFG is enough to accurately capture program behavior and consequently to make accurate performance predictions. For the remainder of this chapter we will thus consider first-order SFGs.

All other program execution characteristics that are included in the statistical profile will be attached to this SFG. For example, consider three instances of basic block A in a statistical profile depending on the previously executed basic block A, B, or C. As such, the program characteristics for A may be different depending on the previously executed basic block. This is to model a *context* in which a basic block is executed.

For each basic block corresponding to a node in the SFG, we record the instruction type of each instruction. We classify the instruction types into 12 classes according to their semantics: load, store, integer conditional branch, floating-point conditional branch, indirect branch, integer alu, integer multiply, integer divide, floating-point alu, floating-point multiply, floating-point divide, and floating-point square root. For each instruction, we record the number of source operands. Note that some instruction types, although classified within the same instruction class, may have a different number of source operands. For each operand we also record the dependency distance, which is the number of dynamically executed instructions between the production of a register value (register write) and the consumption of it (register read). We only consider read-after-write (RAW) dependencies because our focus is on out-of-order architectures in which write-after-write (WAW) and write-after-read (WAR) dependencies are dynamically removed through register renaming as long as enough physical registers are available. Although not done so far, this approach could be extended to also include WAW and WAR dependencies to account for a limited number of physical registers. Note that recording the dependency distance requires storing a distribution, because multiple dynamic versions of the same static instruction could result in multiple dependency distances. In theory, this distribution could be very large because of large dependency distances; in practice, we can limit this distribution. This however limits the number of in-flight instructions that can be modeled during synthetic trace simulation. Limiting the dependency distribution to 512 probabilities allows the modeling of a wide range of current and near-future microprocessors. More formally, the distribution of the dependency distance of the pth operand of the ith instruction in basic block B_n given its context B_{n-1} can be expressed as follows: $\Pr[D_{n,i,p} \mid B_n, B_{n-1}]$.

Note that the characteristics discussed so far are independent of any microarchitecture-specific organization. In other words, these characteristics do not rely on assumptions related to processor issue width, window size, and so on. They are therefore called *microarchitecture-independent characteristics*.

8.2.1.2 Microarchitecture-dependent characteristics

In addition to the preceding characteristics, we also measure a number of characteristics that are related to locality events, such as cache hit/miss and branch predictability behavior. These characteristics are hard to model in a microarchitecture-independent way. Therefore a pragmatic approach is taken, and characteristics for specific branch predictors and specific cache configurations are computed using specialized cache and branch predictor simulators, for example SimpleScalar's sim-bpred and sim-cache [1]. Note that although this approach requires the simulation of the complete program execution for specific branch predictors and specific cache structures, this does not limit its applicability. Indeed, a number of tools exist that measure a wide range of these structures in parallel, for example, the cheetah simulator [22], which is a single-pass, multiple-configuration cache simulator.

The *cache characteristics* consist of the following six probabilities: (1) the L1 instruction cache miss rate, (2) the L2 cache miss rate due to instructions*, (3) the L1 data cache miss rate, (4) the L2 cache miss rate due to data accesses only, (5) the instruction translation lookaside buffer (I-TLB) miss rate, and (6) the data translation lookaside buffer (D-TLB) miss rate.

The *branch characteristics* consist of three probabilities:

1. The probability of a taken branch, which will be used to limit the number of taken branches that are fetched per clock cycle during synthetic trace simulation;
2. The probability of a fetch redirection, which corresponds to a target misprediction—branch target buffer (BTB) miss—in conjunction with a correct taken/not-taken prediction for conditional branches; and
3. The probability of a branch misprediction, which accounts for BTB misses for indirect branches and taken/not-taken mispredictions for conditional branches.

Recall that these microarchitecture-dependent characteristics are measured using specialized cache and predictor simulation, which operates on

---

* We assume a unified L2 cache. However, we make a distinction between L2 cache misses due to instructions and due to data.

an instruction-per-instruction basis. More specifically for the branch characteristics, this means that the outcome of the previous branch is updated before the branch predictor is accessed for the current branch (*immediate update*). In pipelined architectures, however, this situation rarely occurs. Instead, multiple lookups to the branch predictor often occur between the lookup and the update of one particular branch. This is well-known in the literature as *delayed update*. In a conservative microarchitecture, the update occurs at commit time (at the end of the pipeline), whereas the lookup occurs at the beginning of the pipeline by the fetch engine. Delayed update can have a significant impact on overall performance. Therefore, computer architects have proposed speculative update of branch predictors with the predicted branch outcome instead of the resolved outcome. Speculative update can yield significant performance improvements because the branch predictor is updated earlier in the pipeline, for example, at writeback time or at dispatch time. Note that speculative update requires a repair mechanism to recover from corrupted state due to mispredictions. In the results presented in this chapter, we assume speculative update at dispatch time, that is, when instructions are inserted from the instruction fetch queue into the instruction window. It is interesting to note that speculative update mechanisms have been implemented in commercial microprocessors, for example, in the Alpha 21264 microprocessor.

Delayed update, even when using a speculative update mechanism, can have a significant impact on overall performance when modeling microprocessor performance. Therefore branch profiling should take delayed update into account [8]. This can be done using a FIFO (first in first out) buffer in which lookups and updates occur at the head and at the tail of the FIFO, respectively. The branch prediction lookups that are made when instructions enter the FIFO are based on *stale* state that lacks updated information from branch instructions still residing in the FIFO. At each step of the algorithm, an instruction is inserted into the FIFO and removed from the FIFO. A branch predictor lookup occurs when a branch instruction enters the FIFO; an update occurs when a branch instruction leaves the FIFO. If a branch is mispredicted—this is detected upon removal—the instructions residing in the FIFO are squashed and new instructions are inserted until the FIFO is completely filled. In case speculative update is done at dispatch time, a natural choice for the size of the FIFO is the size of the instruction fetch queue. If other update mechanisms are used, such as speculative update at write-back time or nonspeculative update at commit time, appropriate sizes should be chosen for the FIFO buffer.

To show the benefits of the delayed update branch profiling approach, we refer to Figure 8.3, which shows the number of branch mispredictions per 1000 instructions under the following scenarios:

- Execution-driven simulation using SimpleScalar's sim-outorder simulator while assuming delayed update at dispatch time,
- Branch profiling with immediate update after lookup, and
- Branch profiling under delayed branch predictor update.

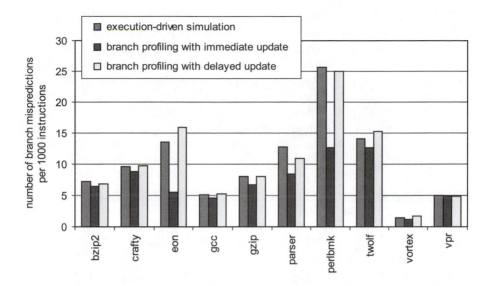

Figure 8.3 The importance of modeling delayed branch predictor update. Reprinted with permission from [8] © 2004 IEEE.

This graph shows that the obtained number of branch mispredictions under immediate branch predictor update can be significantly lower than under execution-driven simulation. Modeling delayed update, however, yields a number of branch mispredictions that is close to what is observed under execution-driven simulation.

8.2.1.3   An example statistical profile

Before moving on with how a synthetic trace is generated from a statistical profile and how this synthetic trace is subsequently simulated, we first give an example of what a statistical profile looks like. Later we consider an excerpt of a statistical profile for basic block A under three different contexts, that is, given the previously executed basic block is A, B, and C. (Note this example is a simplification of a statistical profile that is to be measured from a real program. This example does not show instruction cache misses, nor does it show L2 and TLB misses.)

**A	A**	**A	B**	**A	C**
load	load	load
dep op1: 2 (0.8), 4 (0.2)	dep op1: 2 (0.8), 3 (0.2)	dep op1:3 (0.7), 5 (0.3)
ld miss? yes (1.0)	ld miss? no (1.0)	ld miss? yes (0.6)

add	add	add
dep op1: 1 (1.0)	dep op1: 1 (1.0)	dep op1: 1 (1.0)
dep op2: 4 (0.6), 6 (0.4)	dep op2: 8 (0.5), 6 (0.5)	dep op2: 6 (0.7), 9 (0.3)
sub	sub	sub
dep op1: 1 (1.0)	dep op1: 1 (1.0)	dep op1: 1 (1.0)
dep op2: 2 (1.0)	dep op2: 2 (1.0)	dep op2: 2 (1.0)
br	br	br
dep op1: 3 (1.0)	dep op1: 3 (1.0)	dep op1: 3 (1.0)
taken? no (1.0)	taken? yes (1.0)	taken? yes (0.33)
fetch redirect? (0.01)	fetch redirect? (0.02)	fetch redirect? (0.0)
mispredict? (0.05)	mispredict? (0.02)	mispredict? (0.10)

Obviously, all three instances of basic block A have the same sequence of instructions—namely, *load*, *add*, *sub*, and *branch*—also, the number of inputs to each instruction is equal over all three instances. A dependency that makes two instructions within the same basic block depend on each other (for example, the *add* depends on the *load*) are also the same over all three instances. Dependencies that cross the basic block—an instruction that is dependent on an instruction before the current basic block—can be different for a different context. For example, if the previously executed basic block is A, the *add* instruction has a probability of 60% and 40% to be dependent through its second operand on the fourth and sixth instruction before the *add*; if on the other hand, the previously executed basic block is B, there is a probability of 50% to have a dependency distance of 6 or 8. Similarly for the cache and branch predictor behavior, the characteristic depends on the previously executed basic block. For example, depending on whether the basic block before A is A, B, or C, the probability for a branch misprediction may be different, 5%, 2% and 10% for the preceding example, respectively.

8.2.2   *Synthetic trace generation*

Once a statistical profile is computed, we generate a synthetic trace that is a factor R smaller than the original program execution. R is defined as the *synthetic trace reduction factor*; typical values range from 1000 to 100,000. Before applying our synthetic trace generation algorithm, we first generate a *reduced statistical flow graph*. This reduced SFG differs from the original SFG in that the occurrences of each node are divided by the synthetic trace reduction factor R. In other words, the occurrences in the reduced SFG N_i are a fraction R of the original occurrences M_i for all nodes i:

$$N_i = \left\lfloor \frac{M_i}{R} \right\rfloor.$$

Subsequently, we remove all nodes for which N_i equals zero. Along with this removal, we also remove all incoming and outgoing edges. By doing so, we obtain a reduced statistical flow graph that is no longer fully interconnected. However, the interconnection is still strong enough to allow for accurate performance predictions. Once the reduced statistical flow graph is computed, the synthetic trace is generated using the following algorithm.

1. If the occurrences of all nodes in the reduced SFG are zero, terminate the algorithm. Otherwise, generate a random number in the interval [0,1] to point to a particular node in the reduced SFG. Pointing to a node is done using a cumulative distribution function built up by the occurrences of all nodes. In other words, a node with a higher occurrence will be more likely to be selected than a node with a smaller occurrence.
2. Decrement the occurrence of the selected node reflecting the fact that this node has been accessed. Determine the current basic block corresponding to the node.
3. Assign the instruction types and the number of source operands for each of the instructions in the basic block.
4. For each source operand, determine its dependency distance. This is done using random number generation on the cumulative dependency distance distribution. An instruction x is then made dependent on a preceding instruction $x - d$, with d the dependency distance. Note that we do not generate dependencies that are produced by branches or stores because those types of instructions do not have a destination operand. This is achieved by trying a number of times until a dependency is generated that is not supposedly generated by a branch or a store. If after a maximum number of times (in our case 1000 times) still no valid dependency is created, the dependency is simply squashed.
5. For each load in the synthetic trace, determine whether this load will cause a D-TLB hit/miss, an L1 D-cache hit/miss, and, in case of an L1 D-cache miss, whether this load will cause an L2 cache hit/miss.
6. For the branch terminating the basic block, determine whether this is a taken branch and whether this branch is correctly predicted, results a fetch redirection, or is a branch misprediction.

7. For each instruction, determine whether this instruction will cause an I-TLB hit/miss, an L1 I-cache hit/miss, and, in case of an L1 cache miss, whether this instruction will result in an L2 cache miss.
8. Output the synthetically generated instructions along with their characteristics.
9. If the current node in the reduced SFG does not have outgoing edges, go to step 1, otherwise proceed. Generate a random number in the interval [0,1] and use it to point a particular outgoing edge. This is done using a cumulative distribution built up by the transition probabilities of the outgoing edges. Use this outgoing edge to point to a particular node. Go to step 2.

8.2.3 Synthetic trace simulation

The trace-driven simulation of the synthetic trace is very similar to the trace-driven simulation of real program traces, but the synthetic trace simulator needs to model neither branch predictors nor caches—this is part of the tradeoff that dramatically reduces development and simulation time. However, special actions are needed during synthetic trace simulation for the following cases:

- When a branch is mispredicted in an execution-driven simulator, instructions from an incorrect path are fetched and executed. When the branch gets executed, it is determined whether the branch was mispredicted. In case of a misprediction, the instructions down the pipeline need to be squashed. A similar scenario is implemented in the synthetic trace simulator: When a mispredicted branch is fetched, the pipeline is filled with instructions from the synthetic trace as if they were from the incorrect path; this is to model resource contention. When the branch gets executed, the synthetic instructions down the pipeline are squashed and synthetic instructions are fetched as if they were from the correct path.
- For a load, the latency will be determined by whether this load is an L1 D-cache hit, an L1 D-cache miss, an L2 cache miss, or a D-TLB miss. For example, in case of an L2 miss, the access latency to main memory is assigned.
- In case of an I-cache miss, the fetch engine stops fetching for a number of cycles. The number of cycles is determined by whether the instruction causes an L1 I-cache miss, an L2 cache miss, or a D-TLB miss.

The most important difference between the synthetic trace simulator and the reference execution-driven simulator, other than the fact that the former models the caches and the branch predictor statistically, is that the synthetic trace simulator does not take into account instructions along misspeculated paths when accessing the caches. This can potentially have an impact on the performance prediction accuracy [3].

8.2.4 Simulation speed

Due to its statistical nature, performance metrics obtained through statistical simulation quickly converge to steady-state values. In Eeckhout et al. [8], an experiment was done to quantify the simulation speed of statistical simulation. To this end, the coefficient of variation (CoV) of the instructions per cycle (IPC) was computed as a function of the number of synthetic instructions. The CoV is defined as the standard deviation divided by the mean of the IPC over a number of synthetic traces. The variation that is observed is due to the different random seeds that were used during random number generation for the various synthetic traces—these synthetic traces are different from each other although they exhibit the same execution characteristics. Small CoVs are obtained for small synthetic traces, for example, 4% for 100K, 2% for 200K, 1.5% for 500K and 1% for 1M synthetic instructions. As such, we conclude that synthetic traces containing several hundreds of thousands of synthetically generated instructions are sufficient for obtaining a performance prediction. Note that a synthetic trace of 100K or even 1M synthetically generated instructions is several of orders of magnitude smaller than the (hundreds of) billions of instructions typically observed for real program execution traces. Consequently, statistical simulation allows for simulation speedups by several orders of magnitude compared to full benchmark simulation.

8.2.5 Performance/power prediction accuracy

It is now appropriate to discuss the prediction accuracy of statistical simulation. To this end we first define the *absolute prediction error* for a given metric M in a given design point as follows:

$$AE_M = \frac{M_{statistical\_simulation} - M_{execution-driven\_simulation}}{M_{execution-driven\_simulation}},$$

where $M_{statistical\_simulation}$ and $M_{execution-driven\ simulation}$ is the given metric M in a given design point for statistical and detailed execution-driven simulation, respectively. The metric M could be any metric of interest, for example the IPC, energy consumed per cycle (EPC), the number of used entries in the instruction window, and so on.

Figures 8.4 and 8.5 evaluate the absolute accuracy of statistical simulation for performance prediction and power consumption prediction, respectively. These IPC and EPC numbers are obtained for an eight-wide, out-of-order processor using a framework based on SimpleScalar/Alpha augmented with the Wattch architectural power model using 100M instruction simulation points for a number of Standard Performance Evaluation Cooperative (SPEC) CPU2000 benchmarks—we refer to Eeckhout et al. [8] for a detailed description of the methodology used for obtaining these results. As can be seen from both graphs, the statistical simulation methodology achieves accurate predictions.

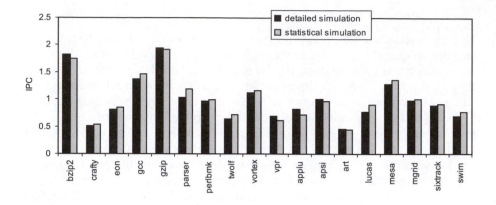

Figure 8.4 Evaluating the absolute performance prediction accuracy of statistical simulation.

For predicting performance or IPC, the average absolute prediction error is 6.9%; for predicting power consumption, the average prediction error is 4.1%. The maximum prediction error that is observed for predicting performance is 16.6%; for predicting power consumption, the maximum prediction error is 12.9%. As such, we conclude that statistical simulation attains fairly accurate performance/power predictions.

In the context of design space explorations, the relative accuracy of a performance model is even more important than the absolute accuracy. A measure of relative accuracy would indicate the ability of a performance estimation technique to predict performance trends, for example, the degree to which performance changes when a microarchitectural parameter is varied. If statistical simulation can provide good relative accuracy, then it can be useful for making design decisions. For example, a designer may want

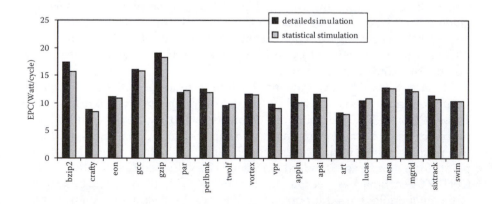

Figure 8.5 Evaluating the absolute power prediction accuracy of statistical simulation.

to know whether the performance gain due to increasing a particular hardware resource justifies the increased hardware cost. Indeed, the sensitivity of power and performance to a particular architectural parameter can help the designer identify the (near) optimal design point, for example, on the "knee" of the performance curve, or where performance begins to saturate as a function of a given architectural parameter.

The *relative prediction error* for a metric M when going from design point A to design point B is defined as:

$$RE_M = \frac{M_{B,\text{statistical\_sim}}/M_{A,\text{statistical\_sim}} - M_{B,\text{execution-driven\_sim}}/-M_{A,\text{execution-driven\_sim}}}{M_{B,\text{execution-driven\_simulation}}/M_{A,\text{execution-driven\_simulation}}}$$

In other words, the relative accuracy quantifies how well statistical simulation is able to predict a relative performance increase or decrease.

Figure 8.6 evaluates the relative accuracy of statistical simulation. Figure 8.6(a) shows performance (IPC) and power consumption as a function of window size, or the number of in-flight instructions; Figure 8.6(b) is a similar

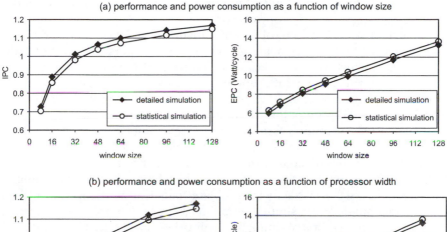

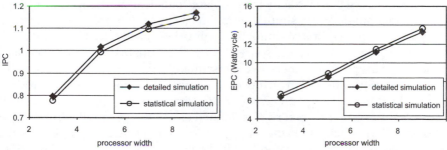

Figure 8.6 Evaluating the relative accuracy of statistical simulation: (a) as a function of window size, and (b) as a function of processor width.

graph showing IPC and power as a function of processor width, or the width of the decode stage, issue stage, and commit stage. Figure 8.6 clearly shows that statistical simulation tracks the performance and power curves very well. The relative error is less than 1.7% in these graphs. A detailed analysis in Eeckhout et al. [8] considering several other microarchitectural parameters and metrics revealed that the relative error for statistical simulation is generally less than 3%.

8.3   Applications

We now discuss a number of interesting applications for statistical simulation: design space exploration, hybrid analytical-statistical modeling, workload space characterization and exploration, program characterization, and system evaluation.

8.3.1   Design space exploration

An important application for statistical simulation is processor design space exploration. Recall that statistical simulation does not aim to replace detailed cycle-accurate simulations. Rather, it aims to provide an efficient look at the design space and to provide guidance for decision making early in the design process. Fast decision making is important to reduce the time-to-market when designing a new microprocessor.

To demonstrate the applicability of statistical simulation for design space explorations, we consider a design space for a superscalar, out-of-order processor in which we vary six microarchitectural parameters: instruction window size, processor width, branch predictor size, L1 instruction cache size, L1 data cache size, and L2 cache size. The total design space consists of 3072 potential design points. In order to evaluate the usefulness of statistical simulation for uniprocessor design space exploration, we need a reference to compare statistical simulation against. We did simulate all 3072 design points through detailed simulation. In order to reduce the total simulation time for doing this (using complete benchmark simulation is impossible to do—note this is exactly the problem statistical simulation addresses), we consider single 100M simulation points as our reference. (See Chapter 7 for a detailed description of simulation points.) We used statistical simulation to explore the same design space. Using the obtained performance and power consumption numbers, we subsequently determine the microarchitectural configuration that achieves the minimum energy-delay product (EDP). EDP is an energy-efficiency metric that is often used in the context of general-purpose processors. It is defined as follows [4]: EDP = EPI $\times$ CPI = EPC $\times$ CPI^2 in which EPI is the energy consumed per instruction, EPC is the energy consumed per cycle and CPI is the number of cycles per instruction. Comparing the optimal architectural configuration as obtained from detailed simulation versus the optimal configuration obtained from statistical simulation, we can determine the error of statistical simulation for design space

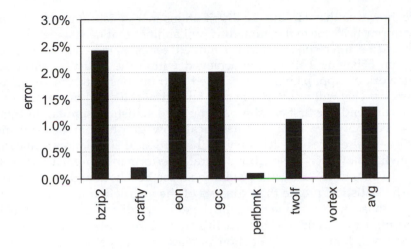

Figure 8.7 Statistical simulation for design space exploration: the error on the minimal-EDP microarchitectural configuration of statistical simulation versus detailed simulation.

exploration. Figure 8.7 shows the error for statistical simulation versus detailed simulation for a number of SPEC CPU2000 benchmarks. The error is 1.3% on average and is no larger than 2.4%, which shows that statistical simulation is indeed capable of identifying a region of optimal design points in a large design space. This region of interesting design points can then be further explored using detailed and thus slower cycle-accurate simulations.

In the preceding experiment we did not consider complete benchmark simulation as our reference but rather used single simulation points. As pointed out in Chapter 7, simulation points reduce the total simulation time significantly compared to complete benchmark simulation. By considering statistical simulation on top of simulation points as done in the above experiment, a 35X simulation speedup is achieved compared to simulation points. Based on the dynamic instruction count of the synthetic trace (1M instructions) versus a simulation point (100M instructions), one might expect a 100X simulation speedup. However, because statistical simulation needs to recompute the statistical profile whenever the cache hierarchy or branch predictor changes during design space exploration, the speedup is limited to 35X, which is still an important simulation speedup. The price paid for this simulation speedup is simulation accuracy, that is, statistical simulation introduces inaccuracies compared to the detailed simulation of simulation points.

8.3.2 Hybrid analytical-statistical modeling

A statistical profile that is used in statistical simulation consists of a number of program characteristics. These characteristics could be varied, and the influence of these parameters on overall performance could be measured.

Note, however, that varying the distributions in a statistical profile is somehow impractical due to the numerous probabilities that need to be specified. It would be interesting to have a limited set of parameters that specify program behavior. This can be achieved within the statistical simulation framework by approximating measured distributions with theoretical distributions. This will result in a hybrid analytical-statistical model.

To this end, we first consider a simplified statistical simulation framework. For now, we omit the statistical flow graph from the statistical profile—this reduces the accuracy of the statistical simulation framework somewhat, however reasonably accurate performance predictions in the range of 10% to 15% can still be obtained, see Eeckhout and De Bosschere [9]. The statistical profile then consists of the instruction mix, the number of operands per instruction type, the dependency distance distribution, the cache miss rates, and the branch misprediction rates averaged over all instructions; that is, no distinction is made per basic block. Looking in more detail on the statistical profile reveals that the instruction mix, the number of operands per instruction type, and the cache and branch predictor characteristics basically are a limited number of probabilities. The dependency distance information, on the other hand, is a distribution consisting of a large number of probabilities, for example, 512. Clearly, approximating the dependency distance distribution by a theoretical distribution (with a limited number of parameters), would result in a compact representation of a program execution. This compact representation then consists of a limited number of program parameters, in the range of 15 to 20 single-value characteristics.

We now study how the dependency distance distribution can be approximated by a theoretical distribution. The probability density function (PDF) of the dependency distance $\Pr[X = x]$ can be written as $\Pr[X = x] = \Pr[X = x \mid X \geq x] \cdot \Pr[X \geq x]$, $x \geq 1$ where $\Pr[X = x \mid X \geq x]$ could be defined as the *conditional dependence probability* $1 - p_x$ (p_x corresponds to the *conditional independence probability* defined by Dubey, Adams, and Flynn [7]); that is, p_x is the probability that an operation is independent on an operation that comes x operations ahead in the instruction trace given that the operation is independent of the $x - 1$ operations ahead of that operation. This equation can be rewritten as follows:

$$\Pr[X = x] = (1 - p_x) \cdot \left(1 - \sum_{i=1}^{x-1} \Pr[X = i]\right).$$

Using induction it can be easily verified that $\Pr[x = x]$ can be written as follows:

$$\Pr[X = x] = (1 - p_x) \cdot \prod_{i=1}^{x-1} p_i, \, x \geq 1.$$

Reverse, calculating p_x from the measured $\Pr[X = x]$ can be done as follows:

$$p_x = 1 - \frac{\Pr[X = x]}{1 - \sum_{i=1}^{x-1} \Pr[X = i]}.$$

Note that any approximation of the conditional independence probability p_x leads to a normalized dependency distance distribution. For example, assuming a conditional independence probability that is independent of x, say p, results in the geometric distribution $\Pr[X = x] = (1 - p) \cdot p^{x-1}$. This approximation was taken by Dubey, Adams, and Flynn [7]. In a follow-up study, Kamin, Adams, and Dubey [14] approximated the conditional independence probability p_x by an exponential function $p_x \approx 1 - \alpha e^{-\beta x}$, where α and β are constants that are determined through regression techniques applied to the measured p_x. More recently, Eeckhout and De Bosschere [9] approximate p_x by a power law: $p_x \approx 1 - \alpha x^{-\beta}$.

Figure 8.8 shows the distribution fitting of the dependency distance distribution for gcc: the conditional dependence probability, the PDF, and the cumulative density function (CDF). The distribution fitting was done by minimizing the sum of squared errors between the theoretical distributions and the measured data of the conditional dependence probability. More specifically, we do the fitting on the raw probability numbers, not on the log-log numbers in these graphs—this gives a higher weight to smaller dependency distances. This is motivated by the fact that this approximation is to be used in an abstract workload model for hybrid analytical-statistical performance modeling, which requires accurate approximations for small dependency distances. Fitting to the conditional dependence probability generally yields more accurate approximations than fitting the probability density function. The graphs in Figure 8.8 show that the power law approximation is more accurate than the exponential approximation for gcc. Similar results are presented in Eeckhout and De Bosschere [9] for other benchmarks. The benchmarks that do not show a nice fit along the power law distribution were programs that spend most of their time in tight loops. For those benchmarks, the dependency distance distribution drops off more quickly than a power law distribution for larger values of x in a log-log diagram.

We can now make use of the power-law properties of the dependency distance distribution to build a hybrid analytical-statistical model. Instead of specifying a distribution consisting of a large number of probabilities, we are now left with only two parameters to characterize the dependency distance characteristic. These two parameters are α and β; α is the probability that an instruction is dependent on its preceding instruction; β is the slope of the conditional dependence probability in a log-log diagram. As a result, the abstract workload model consists of a number of probabilities to characterize the instruction mix, α and β to characterize the interoperation dependencies, and a number of probabilities to characterize the cache hit/miss

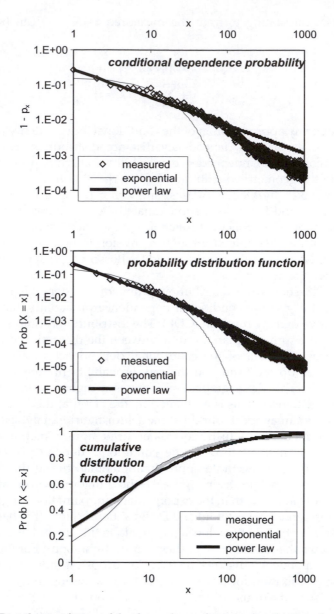

Figure 8.8 Distribution fitting of the dependency distance distribution: the conditional dependence probability, the PDF and the CDF for gcc. Reprinted with permission from [9] © 2001 IEEE.

and branch prediction behavior. This abstract workload model can then be used to drive statistical simulation, yielding a hybrid analytical-statistical modeling approach. Experiments in Eeckhout and De Bosschere [9] show that this hybrid analytical-statistical simulation approach is only slightly less accurate as "classical" statistical simulation using distributions.

8.3.3   Workload space characterization and exploration

As discussed in the previous section, statistical simulation can be used to characterize a program execution by means of an abstract workload model, or a small set of single-value characteristics. This allows for characterizing and exploring the workload space. All the benchmarks can be viewed as points in a multidimensional space in which the various dimensions are the various single-value characteristics.

Figure 8.9 gives an example of a (two-dimensional) workload space characterization for the interoperation dependency characteristics. A number of benchmarks (SPECint95 and IBS, see also Eeckhout and De Bosschere [9]) are shown as a function of α and β from the power law interoperation dependency distance distribution. We can take two interesting conclusions from this graph. First, the interoperation dependencies seem to be quite different for the SPECint95 benchmarks than for the IBS traces. Indeed, all but one of the IBS traces are concentrated in the middle of the graph, whereas the SPECint95 benchmarks are situated more toward the left of the graph. Second, this information can be used to identify weak spots in a workload. For example, this graph reveals there are no benchmarks included for which lies within the interval: 0.28, 0.33. There are two ways to address this lack of benchmark coverage: either search for real benchmarks or generate synthetic traces with the desired program properties. Note that the latter option can be done easily because the program characteristics can be varied freely in a statistical profile. In addition, these program properties can be varied independently from each other.

The important property that the program characteristics can be varied freely and independently from each other within the statistical simulation

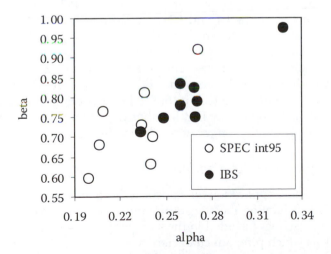

Figure 8.9 Workload space characterization as a function of the dependency distance characteristics α and β. Reprinted with permission from [10] © 2003 IEEE.

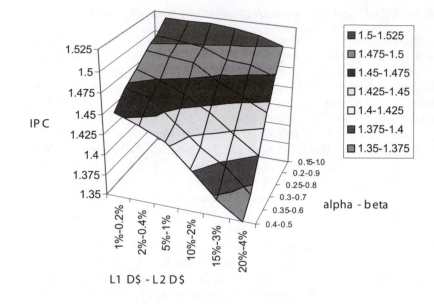

Figure 8.10 Workload space exploration: IPC as a function of data cache miss rates and dependency distance characteristics. Reprinted with permission from [9] © 2001 IEEE.

methodology enables workload space exploration. Workload space explorations are useful because they enable the investigation of the impact of program characteristics and their interactions on overall performance. Such studies are difficult to do using real programs, if not impossible, because most program characteristics and their interaction are hard to vary in real programs. Figure 8.10 gives an example in which IPC is displayed on the z-axis as a function of the L1 and L2 data cache miss rate along the x-axis, and the α and β for the dependency characteristics along the y-axis. This graph clearly shows that the performance of a program with high instruction-level parallelism (ILP) is less affected by the data cache miss rate than a program with low ILP. Indeed, the IPC curve as a function of the data cache miss rate is flatter for ($\alpha = 0.15$; $\beta = 1.0$), denoting high, ILP than for ($\alpha = 0.4$; $\beta = 0.5$), denoting low ILP. In other words, the latency due to cache misses are hidden by the longer dependency distances.

8.3.4   *Program characterization*

Another interesting application for statistical simulation is program characterization. When validating the statistical simulation methodology, in general, and the characteristics included in the statistical profile more, in particular, it becomes clear which program characteristics must be included in the profile for attaining good accuracy. That is, this validation process distinguishes program characteristics that influence performance from those that do not. Several research efforts in the recent past have focused on improving the accuracy of statistical simulation. Improving the accuracy can be achieved by modeling correlation between various program characteristics. For example, Nussbaum and Smith [17] have shown that correlating the instruction mix, the interoperation dependencies, cache miss rates, and branch misprediction rates to the basic block size leads to a significantly higher accuracy in performance prediction. Eeckhout et al. [8] showed the importance of the SFG. Bell et al. [2] showed how the accuracy of statistical simulation improves as the statistical profile evolves from a simple average statistical profile to the SFG as described in this chapter.

8.3.5 System evaluation

Until now, we only discussed uniprocessor performance modeling and the applicability of statistical simulation for addressing the time-consuming simulations. However, for larger systems containing several processors, such as multiprocessors, clusters of computers, and so on, simulation time is even a bigger problem because all the individual components in the system need to be simulated simultaneously. Typically, benchmark problems for such systems are also much larger, and there might be additional design choices. An interesting example is given by Nussbaum and Smith [18] that extends the statistical simulation methodology for evaluating symmetric multiprocessor system performance.

8.4 Previous work

Noonburg and Shen [15] present a framework that models the execution of a program on a particular architecture as a Markov chain, in which the state space is determined by the microarchitecture and in which the transition probabilities are determined by the program execution. This approach was evaluated for in-order architectures. Extending it for wide-resource, out-of-order architectures would result in a far too complex Markov chain.

Hsieh and Pedram [11] present a technique to estimate performance and power consumption of a microarchitecture by measuring a characteristic profile of a program execution, synthesizing a fully functional program from it, and simulating this synthetic program on an execution-driven simulator. The main disadvantage of their approach is the fact that no distinction is made between microarchitecture-dependent and microarchitecture-independent characteristics. All characteristics are microarchitecture-dependent, which makes this technique unusable for design space explorations.

Iyengar et al. [12] present SMART to generate representative synthetic traces based on the concept of a fully qualified basic block. A fully qualified basic block is a basic block together with its context. The *context* of a basic block is determined by its n preceding qualified basic blocks—a qualified basic block is a basic block together with the branching history (of length k) of its preceding branch. This work was later extended in Iyengar et al. [13] to account for cache behavior. In this extended work the focus was shifted from fully qualified basic blocks to fully qualified instructions. The context of a fully qualified instruction is then determined by n singly qualified instructions. A singly qualified instruction is an instruction annotated with its instruction type, its I-cache behavior and, if applicable, its D-cache behavior and its branch behavior. Therefore a distinction is made between two fully qualified instructions having the same preceding instructions, except that, in one case, a preceding instruction missed in the cache, whereas in the other case it did not. Obviously, collecting all these fully qualified instructions results in a huge amount of data to be stored in memory. For some benchmarks, the authors report that the amount of memory needed can exceed the available memory in a machine, so that some information needs to be discarded from the graph. The SFG shares the concept of using a context by qualifying a basic block with its preceding basic block. However, the SFG is both simpler and smaller than the fully qualified graph structure used in SMART. In addition, Eeckhout et al. [8] have found that qualifying with one single basic block is sufficient. Another interesting difference between SMART and the framework presented here is the fact that SMART generates memory addresses during synthetic trace generation. Statistical simulation simply assigns hits and misses.

In recent years, a number of papers [2,6,8,9,10,18,19,20] have been published that are built around (slightly different forms of) the general statistical simulation framework presented in Figure 8.1. The major difference between these approaches is the degree of correlation in the statistical profile. The simplest way to build a statistical profile is to assume that all characteristics are independent from each other [6,9,10], which results in the smallest statistical profile and the fastest convergence time but potentially the largest prediction errors. In HLS, Oskin et al. [18] generate 100 basic blocks of a size determined by a normal distribution over the average size found in the original workload. The basic block branch predictabilities are statistically generated from the overall branch predictability obtained from the original workload. Instructions are assigned to the basic blocks randomly based on the overall instruction mix distribution, in contrast to the *basic block modeling granularity* of the SFG. As in the framework discussed in this chapter, the HLS synthetic trace generator then walks through the graph of instructions. Nussbaum and Smith [18] propose to correlate various characteristics, such as the instruction types, the dependencies, the cache behavior, and the branch behavior to the size of the basic block. Using the size of the basic block to correlate statistics raises the possibility of *basic block size aliasing*, in which statistical distributions from basic blocks with very different characteristics (but of the same size) are combined and reduce simulation accuracy. In a SFG, all characteristics are correlated to the basic block itself, not just its size. Moreover, the SFG correlates basic blocks to its context of previously executed basic blocks; that is, in a first-order SFG, basic blocks with a different previously executed basic block are characterized separately.

8.5 Summary

This chapter discussed statistical simulation of superscalar out-of-order processors. The idea is to measure a well-chosen set of characteristics from a program execution called a *statistical profile*, generate a synthetic trace with those characteristics, and simulate the synthetic trace. If the set of characteristics reflects the key properties of the program's execution behavior, accurate performance/power predictions can be made. The statistically generated synthetic trace is several orders of magnitude smaller than the original program execution, hence simulation finishes very quickly.

The key properties that need to be included in a statistical profile are the statistical flow graph (SFG), the instruction types, the interoperation dependencies, the cache hit/miss behavior, and the branch misprediction behavior. Measuring the branch behavior should consider delayed branch predictor update in order to model delayed update as observed in contemporary microprocessors. The performance and power predictions through statistical simulation are highly accurate: The average absolute error for predicting performance and power is 6.9% and 4.1%, respectively. The relative accuracy is typically less than 3%. Synthetic traces of several hundreds of thousands of instructions are enough to obtain these predictions.

This chapter also discussed five important applications for statistical simulation. For one, design space explorations can be done both efficiently and accurately. Early decision making is important for shortening the time-to-market of newly designed microprocessors. Second, by approximating the distributions contained in a statistical profile using theoretical distributions, the gap between analytical and statistical modeling can be bridged by building a hybrid analytical-statistical model. We focused on the power law properties of the dependency distance characteristics to come to an abstract workload model containing a limited number of single-value workload characteristics. Third, based on such an abstract workload model, workload space characterization and exploration becomes possible. Workload space studies are interesting to compare workloads, uncover weak spots in the workload space, and estimate the impact of program characteristics and their interaction. Fourth, statistical simulation is a useful tool for program characterization, that is, the discrimination of program characteristics that affect performance from those that do not. Fifth, the evaluation of large systems consisting of several microprocessors can also significantly benefit from statistical simulation.

References

1. Austin, T., Larson, E., and Ernst, D., SimpleScalar: An infrastructure for computer system modeling, *IEEE Computer* 35, 2, 59–67, 2002.
2. Bell, R.H. Jr., Eeckhout, L. John, L.K., and De Bosschere, K., Deconstructing and improving statistical simulation in HLS, *Proceedings of the 2004 Workshop on Duplicating, Deconstructing and Debunking*, held in conjunction with ISCA-31, 2–12, June 2004.

3. Bechem, C., Combs, J., Utamaphetai, N., Black, B., Blanton, R.D.S., and Shen, J.-P., An integrated functional performance simulator, *IEEE Micro* 19, 3, 26–35, 1999.
4. Brooks, D., Bose, P., Schuster, S.E., Jacobson, H., Kudva, P.N., Buyuktosunoglu, A., Wellman, J.-D., Zyuban, V., Gupta, M., and Cook, P.W., Power-aware microarchitecture: Design and modeling challenges for next-generation microprocessors, *IEEE Micro* 20, 6, 26–44, 2000.
5. Brooks, D., Tiwari, V., and Martonosi, M., Wattch: A framework for architectural-level power analysis and optimizations, *Proceedings of the 27th Annual International Symposium on Computer Architecture (ISCA-27)*, ACM Press, 83–94, June 2000.
6. Carl, R. and Smith, J.E., Modeling superscalar processors via statistical simulation, *Proceedings of the 1998 Workshop on Performance Analysis and Its Impact on Design*, held in conjunction with ISCA-25, June 1998.
7. Dubey, P.K., Adams, G.B., III, and Flynn, M.J., Instruction window trade-offs and characterization of program parallelism, *IEEE Transactions on Computers*, 43, 4, 431–442, 1994.
8. Eeckhout, L., Bell, R.H., Jr., Stougie, B., De Bosschere, K., and John, L.K., Control Flow Modeling in Statistical Simulation for Accurate and Efficient Processor Design Studies, *Proceedings of the 31st Annual International Symposium on Computer Architecture (ISCA-31)*, 350-361, June 2004.
9. Eeckhout, L. and De Bosschere, K., Hybrid analytical-statistical modeling for efficiently exploring architecture and workload design spaces, *Proceedings of the 2001 International Conference on Parallel Architectures and Compilation Techniques (PACT-2001)*, 25–34, Sept. 2001.
10. Eeckhout, L., Nussbaum, S., Smith, J.E., and De Bosschere, K., Statistical simulation: Adding efficiency to the computer designer's toolbox, *IEEE Micro* 23, 5, 26–38, 2003.
11. Hsieh, C. and Pedram, M., Microprocessor power estimation using profile-driven program synthesis, *IEEE Transactions on Computer-Aided Design* 17, 11, 1080–1089, 1998.
12. Iyengar, V.S. and Trevillyan, L.H., Evaluation and generation of reduced traces for benchmarks, Technical Report RC-20610, IBM Research Division, T.J. Watson Research Center, Oct. 1996.
13. Iyengar V.S., Trevillyan, L.H., and Bose, P., Representative traces for processor models with infinite cache, *Proceedings of the Second International Symposium on High-Performance Computer Architecture (HPCA-2)*, 62–73, Feb. 1996.
14. Kamin, R.A., III, Adams, G.B., III, and Dubey, P.K., Dynamic trace analysis for analytic modelling of superscalar performance, *Performance Evaluation* 19, 2–3, 259–276, 1994.
15. Noonburg, D.B. and Shen, J.P., A framework for statistical modeling of superscalar processor performance, *Proceedings of the Third International Symposium on High Performance Computer Architecture (HPCA-3)*, 298-309, Feb. 1997.
16. Nussbaum, S. and Smith, J.E., Modeling superscalar processors via statistical simulation. *Proceedings of the 2001 International Conference on Parallel Architectures and Compilation Techniques (PACT-2001)*, 15–24, Sept. 2001.
17. Nussbaum, S. and Smith, J.E., Statistical simulation of symmetric multiprocessor systems. *Proceedings of the 35th Annual Simulation Symposium 2002*, 89–97, Apr. 2002.

18. Oskin, M., Chong, F.T., and Farrens, M., HLS: Combining statistical and symbolic simulation to guide microprocessor design, *Proceedings of the 27th Annual International Symposium on Computer Architecture (ISCA-27)*, 71–82, June 2000.

19. Sugumar, R.A. and Abraham, S.G., Efficient simulation of caches under optimal replacement with applications to miss characterization, *Proceedings of the 1993 ACM Conference on Measurement and Modeling of Computer Systems (SIGMETRICS'93)*, 24–35, May 1993.

Chapter Nine

Benchmark Selection

Lieven Eeckhout

Contents

9.1   Introduction

The first step when designing a new microprocessor is to compose a workload that is representative of the set of applications that will be run on the microprocessor when it is used in a commercial product. A workload typically consists of a number of benchmarks with respective input data sets taken from various benchmarks suites, such as SPEC CPU, TPC, MediaBench, and so on (see Chapter 3). This workload will then be used during the various simulation runs to perform design space explorations. It is obvious that *workload design*, or composing a representative workload, is extremely important in order to obtain a microprocessor design that is optimal for the target environment of operation. The question when composing a representative workload is thus twofold: (1) which benchmarks to use and (2) which input data sets to select. In addition, we have to take into account that high-level architectural simulations are extremely time consuming. As such, the total simulation time should be limited as much as possible to limit the time-to-market. This implies that the total number of benchmarks and input data sets should be limited without compromising the final design. Ideally, we would like to have a limited set of benchmark-input pairs spanning the complete workload space, which contains a variety of the most important types of program behavior.

Conceptually, the complete workload design space can be viewed as a p-dimensional space, where p is the number of important program characteristics. Obviously, p will be too large to display the workload design space understandably. In addition, correlation exists between these variables. This reduces the ability to understand what program characteristics are fundamental to make the diversity in the workload space. This chapter presents a methodology to reduce the p-dimensional workload space to a q-dimensional space with $q \ll p$ ($q = 2$ to $q = 4$ typically) making the visualization of the workload space possible without losing important information. This is achieved by using statistical data reduction techniques such as principal components analysis and cluster analysis.

Each benchmark-input pair is a point in this (reduced) q-dimensional space obtained after PCA. We can expect that different benchmarks will be far away from each other, whereas different input data sets for a single benchmark will be clustered together. This representation gives us an excellent opportunity to measure benchmark similarity. Weak clustering indicates dissimilar behavior, whereas strong clustering indicates similar behavior.

Measuring benchmark similarity has multiple interesting applications that will be extensively discussed in this chapter. First, we show how this workload analysis methodology can be used to gain insight into program behavior. We discuss two examples. The first example is studying the impact of input data sets on program behavior. For a single benchmark, weak clustering for different inputs indicates a large impact of the input on overall behavior; strong clustering indicates a small impact. The second example is studying the interaction between a Java application and the virtual machine. Strong clustering for a single virtual machine and different benchmarks indicates overall behavior is largely determined by the virtual machine and not the application—this is observed for short-running Java applications. For longer-running applications we typically observe a weakly clustered behavior for different virtual machines, meaning that the overall behavior is then primarily determined by the application and not the virtual machine.

The second application for the workload analysis methodology discussed in this chapter is workload design, or the selection of a small set of benchmarks that covers the workload space reasonably well. In other words, the reduced workload should be representative of a larger set of benchmarks. This chapter discusses how such a reduced workload can be composed. The basic idea is that if strong clustering is observed in the workload space, it is unnecessary to include all benchmarks from that cluster in the workload; a representative per cluster will suffice. This reduces the total simulation time for two reasons: (1) the total number of benchmark-input pairs is reduced; and (2) we can select the benchmark-input pair with the smallest dynamic instruction count among all the pairs in a cluster.

Our third application is the validation of reduced input sets. KleinOsowski and Lilja [10] proposed so-called *reduced input sets* for the SPEC CPU benchmarks. The goal of reduced input sets is to yield a smaller dynamic instruction count than the reference inputs provided by SPEC while providing a program behavior that is similar to executing the reference input. The workload analysis methodology presented in this chapter can be used to validate reduced inputs.

This chapter is organized as follows. We first discuss general issues related to measuring benchmark similarity. We subsequently detail a workload analysis methodology in section 9.3 that can be used to measure benchmark similarity in a reliable way. This workload analysis methodology is based on two multivariate data analysis techniques, principal components analysis and cluster analysis. Section 9.4 then discusses three important applications: program behavior analysis, workload design, and validation of reduced input sets. Section 9.5 discusses related work, after which we summarize in section 9.6.

9.2   Measuring benchmark similarity

As mentioned in the introduction, the workload space could be viewed as a p-dimensional space in which the dimensions are determined by a set of workload characteristics. The individual benchmarks can then be displayed by a p-dimensional vector within this space. Measuring benchmark similarities within such a multidimensional space poses two important issues that need to be solved.

The first issue is to determine what the dimensions are in this workload space. Intuitively, program characteristics should be used that affect performance. Selecting program characteristics that do not affect performance might discriminate benchmarks on such a characteristic. However, this will not yield information about the behavior of the benchmark when executed on a microprocessor. On the other hand, it is important to incorporate as many program characteristics as possible so that the analysis done on it will be predictive; that is, we want strongly clustered benchmarks to behave similarly. The program characteristics can be obtained from an abstract workload model, from a collection of both microarchitecture-dependent and/or -independent characteristics, from a set of hardware performance monitor values, and from similar sources. Basically, any set of program characteristics can serve as input to this methodology. However, it is up to the user to determine whether the set of program characteristics he or she intends to use is a good choice or not. In this chapter we will consider two data sets: (1) a set of microarchitecture-dependent and -independent program characteristics obtained from running an instrumented binary and (2) a set of hardware performance monitor values obtained from running benchmarks on native hardware. Ideally, we would like to have a concise set of microarchitecture-independent workload characteristics that, when used in conjunction with a performance model, achieves perfect performance predictions over a wide range of microarchitectures. This would allow for a microarchitecture-independent characterization of the workload space. Unfortunately, to date such an analytical model does not exist. By consequence we consider the data sets as mentioned earlier.

The second issue related to measuring benchmark similarity in a multidimensional workload space concerns a metric for quantifying benchmark similarity. A naïve approach would consider the Euclidean distance as a measure for the degree of similarity in the workload space; that is, a small Euclidean distance between two benchmarks suggests similarly behaving benchmarks, whereas a larger Euclidean distance suggests dissimilar program behavior. This naïve approach poses two potential pitfalls. First, in most cases the program characteristics are nonnormalized, meaning that one variable may vary in the range 10 ± 1, whereas another variable may vary in the range 1 ± 0.1. Because of this difference, the first variable has a higher weight in the Euclidean distance than the second variable. For example, a 10% difference along the first variable (which is 1 unit in absolute terms) clearly has a higher weight in the Euclidean distance than a 10% difference in the second variable (which is 0.1 units in absolute terms). This problem can be overcome by normalizing the data set prior to workload analysis. Normalization will transform the data set in such a way that a variable has a zero mean and a unit variance. The second problem related to using the Euclidean distance as a similarity metric in the original workload space is correlation. In case dimensions or variables in a multidimensional space are correlated (or measure a similar underlying program characteristic), a difference between two benchmarks along these correlated dimensions will receive a higher weight than a noncorrelated variable when calculating the Euclidean distance. In the data sets we used in our experiments we observed significant correlation between program characteristics. Removing correlation from a data set can be done using multivariate statistical data analysis techniques as will be discussed in the following section.

9.3 Workload analysis

We first discuss the two data analysis techniques that we use in our workload analysis methodology, namely principal components analysis and cluster analysis. Subsequently, we will detail how these two data analysis techniques can be used in conjunction for workload analysis.

9.3.1 Principal components analysis

Principal components analysis (PCA) [9] is a statistical data analysis technique that presents a different view on a multidimensional data set. It builds on the assumption that many variables (in our case, program characteristics) are correlated and hence measure the same or similar properties of the various cases (in our case, benchmarks). PCA computes new variables called *principal components* that are linear combinations of the original variables, such that all principal components are uncorrelated. PCA transforms the p variables $X_1, X_2, \ldots, X_p$ into p principal components $Z_1, Z_2, \ldots, Z_p$ with

$$Z_i = \sum_{j=1}^{p} a_{ij} \cdot X_j.$$

This transformation has the properties: (1) $\mathrm{Var}[Z_1] \geq \mathrm{Var}[Z_2] \geq \ldots \geq \mathrm{Var}[Z_p]$, which means that Z_1 contains the most information and Z_p the least; and (2) $\mathrm{Cov}[Z_i, Z_j] = 0, \forall i \neq j$, which means that there is no information overlap between the principal components. Note that the total variance in the data remains the same before and after the transformation, namely

$$\sum_{i=1}^{p} X_i = \sum_{i=1}^{p} Z_i \; .$$

Mathematically, PCA actually solves the eigenvalue problem over the correlation matrix.

As stated in the first property in the previous paragraph, some of the principal components have a high variance whereas others have a small variance. By removing the components with the lowest variance from the analysis, we can reduce the number of program characteristics while controlling the amount of information that is thrown away. We retain q principal components, which is a significant information reduction because $q \ll p$ in most cases, typically $q = 2$ to $q = 4$. To measure the fraction of information retained in this q-dimensional space, we use the amount of variance accounted for by these q principal components:

$$\sum_{i=1}^{q} \mathrm{Var}[Z_i] \Big/ \sum_{i=1}^{p} \mathrm{Var}[X_i].$$

Typically 85% to 90% of the total variance should be explained by the retained principal components.

Recall that the p original variables are the program characteristics that build up the original workload space. By examining the most important q principal components, which are linear combinations of the original program characteristics, that is,

$$\left(Z_i = \sum_{j=1}^{p} a_{ij} \cdot X_j, i = 1, \ldots, q \right),$$

meaningful interpretations can be given to these principal components in terms of the original program characteristics. A coefficient a_{ij} that is close to +1 or −1 implies a strong impact of the original characteristic X_j on the principal component Z_i. A coefficient a_{ij} that is close to 0, on the other hand, implies no impact.

The next step in the analysis is to display the various cases or benchmarks as points in the q-dimensional space built up by the q principal components. This can be done by computing the values of the q retained principal components for each benchmark. As such, a view can be given on the workload space, and benchmark similarity can be studied. The projection in the q-dimensional space is much easier to understand than a view of the original p-dimensional space for two reasons: (1) q is much smaller than p: $q << p$, and (2) the q-dimensional space is uncorrelated.

During PCA, one can work either with normalized or nonnormalized data (the data is normalized when the mean of each variable is zero and its variance is one). In the case of nonnormalized data, a higher weight is given in the analysis to variables with a higher variance. As mentioned before, we have used normalized data because of our heterogeneous data sets; for example, the variance of variable x can be orders of magnitude larger than the variance of another variable y.

9.3.2   *Cluster analaysis*

Cluster analysis (CA) [9] is another data analysis technique that is aimed at clustering n cases, in our case benchmarks, based on the measurements of p variables, in our case program characteristics. The final goal is to obtain a number of groups, containing various benchmarks that exhibit similar behavior. There exist two commonly used types of clustering techniques, namely linkage clustering and K-means clustering.

Linkage clustering starts with a matrix of distances between the n cases or benchmarks. As a starting point for the algorithm, each case is considered a group. In each iteration of the algorithm, the two groups that are closest to each other (with the smallest distance in the p-dimensional space, also called the *linkage distance*) will be combined to form a new group. As such, close groups are gradually merged until finally all cases are in a single group. This can be represented in a so-called *dendrogram*, which graphically represents the linkage distance for each group merge in each iteration of the algorithm. Having obtained a dendrogram, it is up to the user to decide how many clusters to take. This decision can be made based on the linkage distance. Indeed, small linkage distances imply strong clustering whereas large linkage distances imply weak clustering. There exist several methods for calculating the distance between groups or clusters of cases all potentially leading to different clustering results. In this chapter, we consider two possibilities: the *furthest neighbor* strategy (also known as *complete linkage*) and the *weighted pair-group average* strategy. In complete linkage, the distance between two clusters is computed as the largest distance between any two cases from the clusters (or thus the furthest neighbor). In the weighted pair-group average method, the distance between two clusters is computed as the weighted average distance between all pairs of cases in two different clusters. The weighting of the average is done by considering the cluster size, that is, by taking into account the number of cases per cluster.

K-means clustering produces exactly K clusters with the greatest possible distinction—Chapter 7 on SimPoint also considers K-means clustering for computing simulation points. The K-means clustering algorithm works as follows. In each iteration, the distance is calculated for each case to the center of each cluster. A case then gets assigned to the closest cluster. As such, new clusters are formed and new cluster centers can be computed. This algorithm is iterated until no more changes are observed. It is well-known that the result of K-means clustering can be dependent on the choice of the initial cluster centers. SimPoint (see Chapter 7) solves this issue by considering different randomly chosen initial cluster centers and picking the one that results in the best clustering result. In this chapter we use a different approach by maximizing the distance between the various initial cluster centers as an initial estimate.

9.3.3 Putting it together

The workload analysis methodology presented in this chapter combines PCA and CA and consists of the following steps:

1. The p program characteristics of interest are measured for a set of n benchmarks. As mentioned earlier, any data set of interest can be used as input for this workload analysis methodology. However, it is up to the user to determine whether the chosen program characteristics is appropriate or not.
2. In a second step, these p program characteristics are normalized over all n benchmarks so that for each program characteristic, the average equals zero and the variance equals one. Note that depending on the goal of the workload analysis and the given data set, it might be undesirable to normalize. In our experiments however, normalizing the data prior to analysis was necessary.
3. On these normalized data points, PCA is performed. This can be done using an existing statistical software package; in all our experiments we used STATISTICA [16], a commercial software package for statistical computations. This is done by presenting a two-dimensional matrix as input to the statistical software in which the columns are the variables (program characteristics) and the rows the cases (benchmarks). PCA is then performed by the software, which yields p principal components.
4. Now, it is up to the user to determine how many principal components need to be retained. We will denote the number of retained principal components as q. This decision can be made based on the

amount of variance accounted for by the retained principal components. For example, one could consider all principal components that have a variance that is larger than the lowest variance prior to PCA—in case normalized variables are used, all principal components are retained that have a variance larger than one. One could also retain as many principal components to explain a given percentage of the total variance, for example 85%, 90%, or 95%.

5. The q retained principal components can be analyzed, and a meaningful interpretation can be given to them. This is done based on the coefficients a_{ij}, also called the *factor loadings*, as they occur in the following equation:

$$Z_i = \sum_{j=1}^{p} a_{ij} \cdot X_j, \text{ with } Z_i, 1 \le i \le q, \text{ the retained principal components}$$

and $X_j, 1 \le j \le p$, the original program characteristics
A positive coefficient a_{ij} means a positive impact of program characteristic X_j on principal component Z_i; a negative coefficient a_{ij} implies a negative impact. If a coefficient a_{ij} is close to zero, this means X_j has (nearly) no impact on Z_i.

6. The benchmarks can be displayed in the workload space built up by these q principal components. This is done by computing a q-dimensional vector per benchmark built up by the retained principal components:

$$Z_i = \sum_{j=1}^{p} a_{ij} \cdot X_j, 1 \le i \le q$$

Each benchmark is then plotted in this q-dimensional space.

7. Rescale the q principal components to unit variance.

8. Within this rescaled q-dimensional space the Euclidean distance can be computed between the various benchmarks as a reliable measure for the way benchmarks differ from each other. There are two reasons supporting this statement. First, the values along the axes in this space are uncorrelated because they are determined by the principal components, which are uncorrelated by construction. Second, by rescaling the principal components (previous step), the principal components are placed on a common scale. Without rescaling, the variance of a principal component—which is a manifestation of the correlation in the original data—would give a higher weight in the calculation of the Euclidean distance to correlated characteristics in the original data.

9. Finally, CA can be done within this rescaled q-dimensional space. The Euclidean distance gives a reliable measure for benchmark similarity. The dendrogram obtained from cluster analysis gives a clear view on the (dis)similarity between the various benchmarks.

An alternative to applying PCA prior to cluster analysis is to use the Mahalanobis distance* in the original workload space as the similarity metric for driving the cluster analysis. However, we see two advantages of using PCA. First, PCA gives us the opportunity to visualize the workload space in an understandable way because the number of dimensions in the transformed workload space is limited. This facilitates reasoning and analyzing the workload space. Second, PCA helps us in explaining why benchmarks differ from each other in terms of the original program characteristics. Note, however, that the Mahalanobis distance is equivalent to the distance measure that is used in this paper if all the principal components would have been used in our calculation of the Euclidean distance. Because we keep only the leading principal components, which account for all but a small fraction of the total variance, our distance measure becomes a very close approximation of the real Mahalanobis distance.

9.4 Applications

We now discuss three important applications for this workload analysis methodology: program behavior analysis, workload design, and the validation of reduced input sets.

9.4.1 Program Behavior Analysis

Our first application is program behavior analysis. This is done using two examples. The first example shows studying the impact of input data sets on overall program behavior [3,4]. The second example is investigating the interaction between Java applications and virtual machines [2].

9.4.1.1 Impact of input data sets

Table 9.1 shows the program characteristics that are used to study the impact of input data sets on program behavior. There are 18 ($p = 18$) program characteristics in total covering both microarchitecture-dependent and –independent characteristics. These characteristics cover the instruction mix, branch predictability, control flow, data cache behavior, instruction cache behavior, and inherent instruction-level parallelism (ILP). These program characteristics were obtained for a number of benchmarks through instrumentation. The instrumentation was done using ATOM [15], a binary instrumentation tool for

---

* The Mahalanobis distance incorporates correlation between the various dimensions in its metric. For calculating the Mahalanobis distance, the overall covariance matrix of the original characteristics X_i could be used.

Table 9.1 A Set of Microarchitecture-Dependent and Independent Program
Characteristics to Drive the PCA Analysis

No.	Category	Program Characteristic
1	Instruction mix	Percentage integer arithmetic operations
2		Percentage logical operations
3		Percentage shift and byte manipulation operations
4		Percentage load/store operations
5		Percentage control operations
6	Branch predictability	Branch prediction accuracy for a hybrid branch predictor selecting among an 8K-entry bimodal predictor and an 8K-entry gshare predictor (history of 12 branches); meta predictor contains 8K entries
7	Control flow	Number of instructions between two sequential flow breaks, or the number of instructions between two taken branches
8	Data stream behavior	Miss rate for a 8KB direct-mapped D-cache
9		Miss rate for a 16KB direct-mapped D-cache
10		Miss rate for a 32KB 2-way set-associative D-cache
11		Miss rate for a 64KB 2-way set-associative D-cache
12		Miss rate for a 128KB 4-way set-associative D-cache
13	Instruction stream	Miss rate for a 8KB direct-mapped I-cache
14	behavior	Miss rate for a 16KB direct-mapped I-cache
15		Miss rate for a 32KB 2-way set-associative I-cache
16		Miss rate for a 64KB 2-way set-associative I-cache
17		Miss rate for a 128KB 4-way set-associative I-cache
18	Instruction-level parallelism (ILP)	ILP on an infinite-resource processors, i.e. assuming an infinite number of functional units, infinite decode/issue/reorder width, infinite window size, perfect caches, perfect branch prediction, unit execution latency. In other words, only read-after-write dependencies are considered through registers as well as through memory.

the Alpha architecture. ATOM instruments a binary that when run on native hardware collects program execution characteristics. ATOM is capable of instrumenting functions, basic blocks, and individual instructions.

In this study, we consider the SPECint95 benchmarks* in conjunction with a database workload consisting of TPC-D queries**. The reason why we chose SPECint95 instead of the more recent SPECint2000 is to limit the simulation time. SPEC opted to dramatically increase the runtimes of the SPEC2000 benchmarks compared to the SPEC95 benchmarks, which is beneficial for performance evaluation on real hardware but impractical for simulation purposes. In addition, SPECint95 has more reference inputs per benchmark than

* http://www.spec.org
** http://www.tpc.org

SPECint2000. For gcc (GNU C compiler) and li (lisp interpreter), we used all the reference input files: 26 inputs for gcc and 12 inputs for li. For ijpeg (image processing, we used the three reference inputs along with 8 other input images taken from the Web; the images were chosen in such a way to cover a broad range of resolutions ranging from 512 × 480 pixels up to 2362 × 1570 pixels. For compress (text compression), we have adapted the reference input "14,000,000 e 2231" to obtain different input sets: "100,000 e 2231", "500,000 e 2231", "1,000,000 e 2231", "5,000,000 e 2231" and "10,000,000 e 2231". For m88ksim (microprocessor simulation) and vortex (object-oriented database), we have used the train and the reference inputs as provided in the SPEC CPU distribution. The same was done for perl (perl interpreter): jumble was taken from the train input, and primes and scrabbl were taken from the reference input as well as for go (game)—"50 9 2stone9.in" from the train input, and "50 21 9stone21.in" and "50 21 5 stone21.in" from the reference input.

In addition to SPECint95, we used postgres v6.3 running the decision support TPC-D queries over a 100-megabyte Btree-indexed database. For postgres, we ran all TPC-D queries except for query 1 because of memory constraints on our machine.

The benchmarks were compiled with optimization level -O4 and linked statically with the -non_shared flag for the Alpha architecture. In total we thus have 79 program-input pairs.

Figure 9.1, which is a result of applying PCA on this 79 (number of program-input pairs) × 18 (program characteristics) matrix, shows that 4

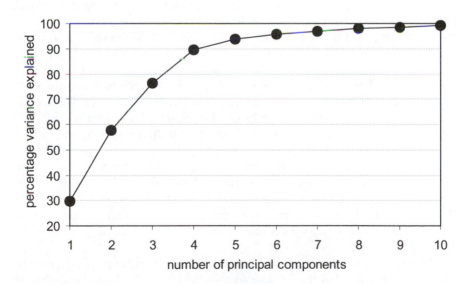

Figure 9.1 Amount of variance explained as a function of the number of principal components.

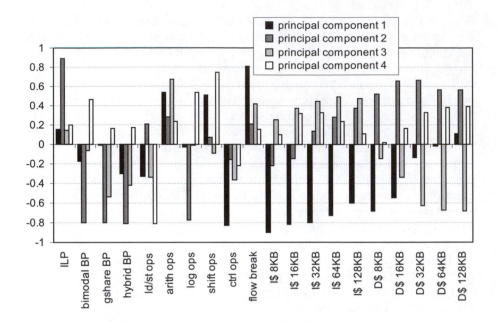

Figure 9.2 Factor loadings.

($q = 4$) principal components are sufficient to account for 89.5% of the total variance. Intuitively speaking, these four principal components account for 89.5% of the information contained in the original data set. The first component accounts for 29.7% of the total variance and is positively dominated, see Figure 9.2, by the number of instructions between two taken branches; and negatively dominated by the percentage control operations and the I-cache miss rates. This means that program-input pairs with a high value along the first principal component typically have a relatively large number of instructions between two taken branches and a relatively low number of control transfer instructions and relatively low I-cache miss rates. The second principal component accounts for 28% of the total variance and is positively dominated by the amount of ILP and negatively dominated by the branch prediction accuracy and the percentage of logical operations. The third component accounts for 18.5% of the total variance and is positively dominated by the percentage arithmetic operations and negatively dominated by the D-cache miss rates for large cache sizes. The fourth component accounts for 13.3% of the total variance and is positively dominated by the percentage shift operations and negatively dominated by the percentage load/store operations.

Based on these results, we can now compute the Principal Components (PC) values for each of the 79 program-input pairs. Plotting these 79 program-input pairs in the (rescaled) PCA space gives a workload space view as given in Figure 9.3. The various program-input pairs are plotted in a

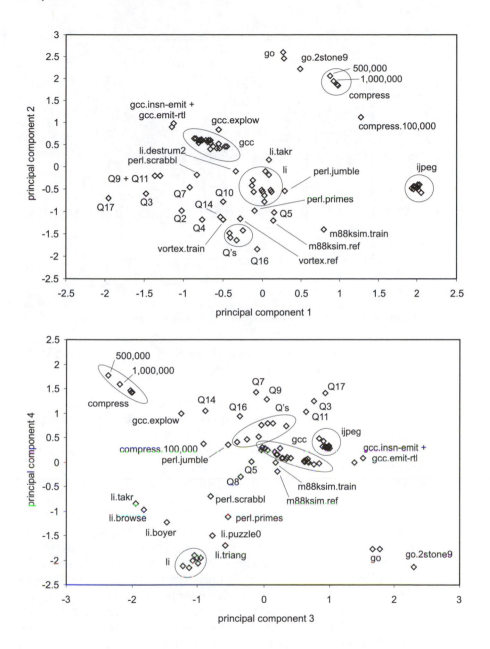

Figure 9.3 Workload space after PCA: PC_1 versus PC_2 in the top graph and PC_3 versus PC_4 in the bottom graph.

four-dimensional space by plotting PC_1 as a function of PC_2 and PC_3 as a function of PC_4. As mentioned earlier, CA can be applied on this rescaled PCA space. The dendrogram obtained from cluster analysis is shown in Figure 9.4. Based on the workload space view given in Figure 9.3 and the

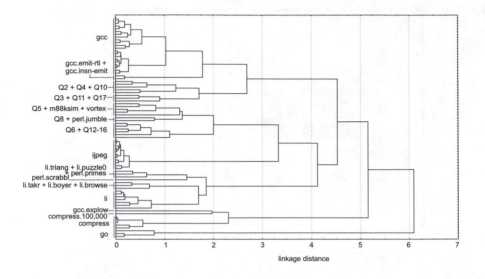

Figure 9.4 Dendrogram obtained from complete linkage cluster analysis.

dendrogram given in Figure 9.4, several interesting conclusions can be taken, as they are discussed next.

Isolated Benchmark Behavior. Apparently, go, ijpeg, and compress are isolated in this four-dimensional space. Indeed, in the dendrogram shown in Figure 9.4, these three benchmarks are connected to the other benchmarks through long linkage distances. For example, go is connected to the other benchmarks with a linkage distance of 4.6, which is much larger than the linkage distance for more strongly clustered benchmark-input pairs. An explanation for this phenomenon can be found in Figure 9.3. For go the discrimination is made along the second and third component. This is due to its low branch prediction accuracy, its low percentage logical operations, its high amount of ILP, its high percentage of arithmetic operations, and its low D-cache miss rates for larger cache sizes. Compress discriminates itself along the third component, which is mainly due to its high D-cache miss rates for large caches. For ijpeg, the different behavior is due, along the first and fourth component, to the high percentage of arithmetic, shift and control operations, the high number of instructions between two taken branches, the low percentage of load/store and control operations, and the low I-cache miss rates.

Strong Clustering. There are also several strong clusters containing several input sets for the same benchmark. This suggests that the input set has a minor impact on overall performance for those benchmarks. An important consequence of this observation is that, as will be discussed later on in this chapter, only a small number (or in some cases, only one) of the input sets should be selected to represent the whole cluster. This will ultimately reduce the total simulation time because only a few (or only one) program-input pairs need to be simulated instead of all the pairs within that cluster. We can identify several strong clusters:

- The data points corresponding to gcc are strongly clustered, except for inputs emit-rtl, insn-emit, and explow. These three inputs exhibit a different behavior from the rest of gcc's inputs. Nevertheless, emit-rtl and insn-emit are quite similar to each other.
- The data points corresponding to the lisp interpreter li, except for browse, boyer, takr, triang, and puzzle0 are strongly clustered as well.
- All input sets for ijpeg result in similar program behavior because all inputs are clustered in one group. An important conclusion from this analysis is that the behavior of ijpeg is unaffected by the image dimensions.
- The inputs for compress are strongly clustered as well except for the smallest input "100,000 e 2231".

Weak clustering. The data points corresponding to postgres running the TPC-D queries are weakly clustered. For example, the spread along the first principal component is very large. As such, a wide range of different I-cache behavior can be observed when running different TPC-D queries, that is, a different query seems to result in a different instruction footprint. Also for perl we observe a large impact of the input of overall program behavior; all three inputs result in significantly different program behavior.

Reference versus train inputs. Along with the CPU benchmark suite, SPEC releases both reference and train inputs. The purpose for the train inputs is to provide input sets that should be used for profile-based compiler optimizations. The reference input is then used for reporting results. Within the context of this chapter, the availability of reference and train input sets is important for two reasons. First, when reference and train inputs result in similar program behavior we can expect that profile-driven optimizations will be effective. Second, train inputs have smaller dynamic instruction counts, which make them candidates for more efficient simulation runs. In other words, when a train input exhibits a similar behavior as a reference input, the train input can be used instead of the reference input for exploring the design space, which will lead to a more efficient design flow.

In this respect, we draw the following conclusions:

- The train and reference input for vortex exhibit similar program behavior with a linkage distance that is smaller than 0.4.
- For m88ksim on the other hand, this is less the case—the linkage distance is larger than 1.
- For go, the train input "50 9 2stone9.in" leads to a behavior that is slightly different from the behavior of the reference inputs

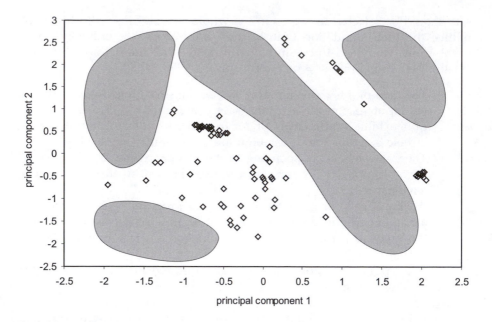

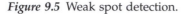

Figure 9.5 Weak spot detection.

"50 21 9stone21.in" and "50 21 5stone21.in". The two
reference inputs on the other hand, lead to similar behavior.

From these observations, we can state that for some benchmarks the
train input behaves similarly to the reference input. For other benchmarks
this might not be true. As such, using train inputs when reporting performance results in architectural research might be reliable in some cases and
unreliable in other cases.

Weak Spot Detection. Another interesting insight that can be obtained
from applying PCA is whether the workload covers the workload space well.
Figure 9.5 again shows the workload space along the first two principal
components; the weak spots are highlighted through gray shapes. The identification of weak spots is valuable information for performance analysts and
computer architects as they try to compose a benchmark suite that covers
the complete workload space. A weak spot can then be addressed by searching for additional benchmarks or by using synthetic workloads with properties matching the weak spot characteristics. An example of how synthetic
workloads can be generated are discussed in the previous chapter.

9.4.1.2 Java workloads

In our second example of how the workload analysis methodology presented
in this chapter can be used for gaining insight in program behavior, we consider
Java workloads that are executed within a managed run-time environment or

a virtual machine (VM). The purpose of this study [1] was to better understand the interaction between the VM and the Java application being executed. Moreover, we wanted to gain insight on how overall Java application
behavior is affected by the VM, the Java application itself, and the input to
the Java application.

This research was motivated by the observation that previous work on
Java workload characterization typically considered only one or two VMs in
their methodology as well as only one benchmark suite, mostly SPECjvm98.
In addition, some studies use a small input set, for example, s1 for
SPECjvm98, to limit the simulation time in their study. As such, we can raise
several relevant questions in relation to this methodology. Is such a methodology reliable for Java workloads? What if the behavior of a Java workload
is highly dependent on the chosen VM? Can we translate conclusions made
for one VM to another? Is using a small input, for example, SPECjvm98's
s1, yielding a short-running Java workload representative for a large input,
such as s100?

For this study, we considered seven VMs: SUN JRE 1.4.1, Blackdown
JRE 1.4.1, IBM JRE 1.4.1, JikesRVM baseline, JikesRVM adaptive, BEA
Weblogic's JRockit, and Kaffe. We considered the SPECjvm98 benchmarks.
(In Chow et al. [1], we considered additional benchmarks; however, the
results for those will not be discussed in this chapter. We refer the interested
reader to Chow et al. [1] for more details.) In this section, we consider
microarchitecture-dependent program characteristics as measured through
hardware performance monitors. Hardware performance monitors are
microprocessor-specific registers that can measure a wide variety of performance events during the native hardware execution of a computer program.
We refer to Chapters 11, 12, and 13 for an extensive discussion on hardware
performance monitors. In the study discussed here, the performance monitors were measured on an AMD K7 Duron microprocessor. We measure 33
events through hardware performance monitors.

Figures 9.6 and 9.7 show the rescaled PCA space for SPECjvm98's s1
and s100 input sets, respectively. These figures only show the first two
dimensions of these workload spaces. For the SPECjvm98 s1 data set, the
PCA analysis revealed four PCs are needed to account for 86.5% of the total
variance; for the SPECjvm87 s100 data set, six PCs are needed to account for
87.3% of the total variance. Figures 9.6 and 9.7 only show the first two
dimensions; the other dimensions did not show any additional information.

Figure 9.6 shows that for the s1 data set the data points are more or less
clustered per VM. From these results we can thus conclude that, for the s1
input set, the VM has a larger impact on the overall behavior than the Java
application itself. In other words, a VM running a Java application with a
small input will exhibit similar behavior irrespective of the Java application
it is running. This can be understood intuitively because the s1 input set
results in very short-running benchmarks (in the order of seconds) for which
the startup time of the VM (initializing and loading parts of the JDK library)
is the highest factor contributing to the overall behavior. From these data

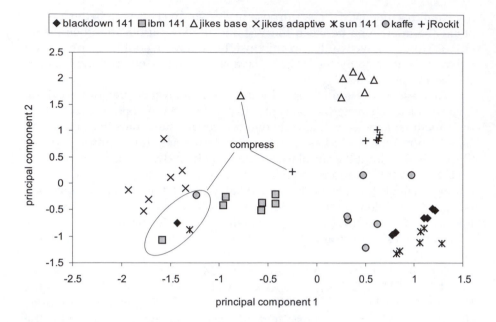

Figure 9.6 PCA space for seven virtual machines and the SPECjvm98 benchmarks with the s1 input data set.

we can also conclude that using the s1 input set of SPECjvm98 in a performance analysis might not be a good method unless one is primarily interested in measuring startup times, not just long-running performance.

It is also interesting to note that the data points corresponding to the compress benchmark are not part of the clusters discussed earlier. In other words, for this Java benchmark, the interaction between the application and the VM has a large impact on its overall behavior at the microarchitectural level because the various VMs running compress are spread over the Java workload space. A close inspection of compress reveals that it has a small code size, while processing a fairly large amount of data, even in case of the s1 input set. Profiling shows that for this benchmark, the top 10 methods account for 98% of all method calls. Compress thus has a small number of hot methods, much smaller than the other SPECjvm98 benchmarks. This leads to a small working set and allows fairly aggressive optimizations by the VM's native code generator. Because each VM implements its run-time optimizer in a different way, this can result in a behavior that is quite different for each VM.

Referring now to Figure 9.7, which shows the rescaled PCA space for the SPECjvm98 s100 data set, it is obvious that the clusters are no longer formed around VMs, as is the case for the s1 input set. For the s100 input set, we observe *benchmark clusters*—the same benchmark being run on different VMs, or small impact of VM on overall behavior—as well as *virtual*

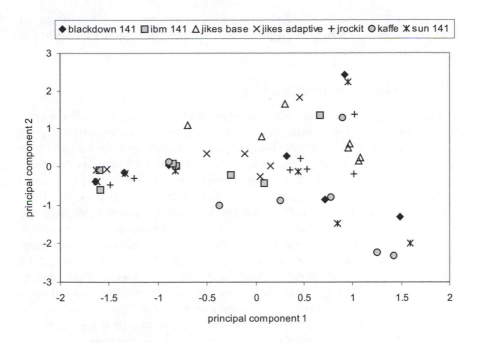

Figure 9.7 PCA space for seven virtual machines and the SPECjvm98 benchmarks with s100 input data set.

machine clusters—the same VM running different Java applications, or large impact of VM on overall behavior. In Figure 9.7, we observe three tight benchmark clusters: (1) a cluster corresponding to compress, (2) a cluster corresponding to mpegaudio, and (3) a cluster corresponding to db. The first two clusters contain all the VMs except for the baseline version of Jikes. The last cluster around db contains five VMs, all but Kaffe and the baseline version of Jikes. Interestingly, Shuf et al. [14] labeled these SPECjvm98 benchmarks as "simple" benchmarks. The fact that the VMs running these "simple" benchmarks result in clustered data points is probably (and surprisingly) due to the fact that all the VMs have optimized these simple benchmarks to nearly the same native code during the long-running time of these benchmarks. Note that in contrast to the widespread behavior of compress for the s1 input, the s100 input results in a tight cluster.

In addition to these three benchmark clusters, we observe two tight VM clusters: (4) the baseline version of the Jikes VM, and (5) the JRockit VM. The cluster around the baseline Jikes VM contains all the SPECjvm98 benchmarks. The fact that the various Java programs that are run on baseline Jikes exhibit similar behavior can be explained as follows. The baseline configuration of Jikes compiles each method just in time, but the number of (dynamic) optimizations performed is limited. As such, we can expect that more or less the same code sequences will be generated for different Java programs yielding similar behavior. The cluster around JRockit contains all the SPECjvm98 benchmarks except for the "simple" benchmarks, compress, db, and mpegaudio.

9.4.2   Workload composition

We now go back to the data set of section 9.4.1.1, that is, the collection of microarchitecture-dependent and –independent characteristics for 79 SPECint95 and TPC-D benchmark-input pairs. Assume for now that all these 79 program-input pairs are representative for the target domain of operation; that is, ideally we would like to consider all these benchmark-input pairs throughout the microprocessor design process. Obviously, simulating these 79 benchmark-input pairs on an architectural simulator is infeasible. Indeed, the total dynamic instruction count of all these program-input pairs together exceeds 593 billions of instructions, or 23 days of simulation when using SimpleScalar's out-of-order simulator at a typical simulation speed of 300,000 instructions per second. As such, it would take 3 weeks of simulation for obtaining performance metrics for one single microarchitectural design point. If we take into account that a large number of design points need to be evaluated, we can conclude that this approach is impractical. One possible solution to this problem would be to run simulations in parallel on a cluster of machines. Because machines are quite cheap nowadays, the equipment cost can be modest. However, the simulations might still be too time-consuming because the total simulation time on a cluster of machines is as long as the slowest simulation. For example, simulating one single microarchitectural configuration for vortex-ref with a dynamic instruction count of more than 92 billion instructions, still takes 3.5 days. Thus, the total simulation time for a complete benchmark suite cannot be faster than 3.5 days.

To address this issue, we could pick a limited number of benchmark-input pairs from the workload space. These benchmark-suite pairs should be chosen in such a way that they are both representative and have a small dynamic instruction count. In this section we show how the workload analysis methodology from this chapter can be used to reduce a workload with a large number of program-input pairs to a set with a more limited set. In section 3.3, we have discussed three possible clustering strategies: (1) linkage clustering using the complete linkage rule, (2) linkage clustering using the weighted pair-group average linkage rule, and (3) K-means clustering. All three clustering mechanisms can be used for selecting a representative workload. Using K-means clustering for this purpose is straightforward. If the number of benchmark-input pairs should be limited to 16 benchmarks, the variable K needs to set to 16. K-means clustering then determines 16 clusters that best fit the data set; for each cluster a representative then needs to be chosen. Linkage clustering techniques can also be used for the purpose of representative benchmark selection. We will now discuss how this can be done. Figure 9.8 shows the dendrogram obtained from linkage clustering using the complete linkage rule—Figure 9.8 is basically the same as Figure 9.4. The vertical bold dashed line shows the linkage

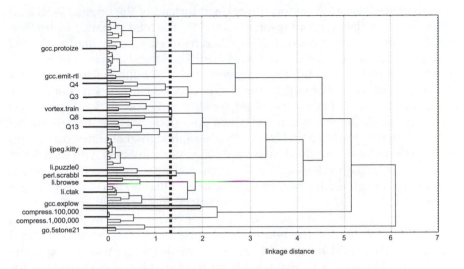

Figure 9.8 Workload design through complete linkage cluster analysis.

distance at which 16 clusters are defined. For each cluster, we can now determine a representative. The representative for each cluster was chosen by taking the program-input pair with the minimal dynamic instruction count that is as close as possible to the center of the cluster it belongs to. The results are shown in Table 9.2: The clusters are shown along with their

Table 9.2 Workload Design Using Complete Linkage Cluster Analysis

Representative	Benchmarks
postgres.Q4	postgres.Q2, Q4, Q7, Q9, and Q10
postgres.Q13	postgres.Q6, Q12 to Q16
postgres.Q3	postgres.Q3, Q11 and Q17
vortex.train	postgres.Q5, m88ksim (ref and train) and vortex (ref and train)
postgres.Q8	postgres.Q8 and perl.jumble
gcc.protoize	all inputs for gcc except for insn-emit, emit-rtl and explow
gcc.emit-rtl	gcc.emit-rtl and gcc.insn-emit
gcc.explow	gcc.explow
li.ctak	all inputs for li except for takr, browse, boyer, triang and puzzle0
li.browse	li.takr, li.browse and li.boyer
li.puzzle0	li.triang, li.puzzle0 and perl.primes
compress.1,000,000	all inputs for compress except for compress.100,00
compress.100,000	compress.100,000
ijpeg.kitty	all inputs for ijpeg
perl.scrabbl	perl.scrabbl
go.5stone21	all inputs for go

representative. Using this reduced workload instead of the original workload results in a simulation speedup of a factor 7.8—the total dynamic instruction count for the reduced workload is 76 billion instructions compared to the 593 billion instructions for the original workload. The largest dynamic instruction count that is observed in this reduced workload is 35 billion instructions. This is nearly a factor 2.6 shorter than the largest dynamic instruction count in the original workload. In Eeckhout et al. [4], we also studied workload composition using linkage clustering with the weighted pair-group average linkage rule, as well as *K*-means clustering, and concluded that although small differences might occur in the reduced workloads, the results are quite consistent with each other.

In the approach discussed earlier for selecting a representative from a cluster, we considered the benchmark-input pair that is closest to the cluster center. Another approach would be to pick a limited number of extreme program-input pairs from each cluster—an extreme point in a cluster is a point that is situated at the boundary of the cluster. The rationale behind this approach would be that the behavior of program-input pairs in the middle of a cluster can be extracted from the behavior of the extremes, for example, through interpolation. A potential problem with this approach is that processor performance of a program-input pair in the middle of a cluster cannot be accurately estimated by interpolating between extremes because determining the interpolation curve is extremely difficult. The reason is that the influence of a program characteristic in one processor configuration can be completely different from the influence in case of another processor configuration. For this reason, we suggest selecting a representative that is close enough to the center of its cluster.

9.4.3   Reduced input sets

KleinOsowski and Lilja [10] propose to reduce the simulation time of the SPEC CPU2000 benchmark suite by using reduced input data sets, called MinneSPEC.* These reduced input sets are derived from the reference inputs by a number of techniques: modifying inputs (for example, reducing the number of iterations), truncating inputs, and so on. The benefit of these reduced inputs is that the dynamic instruction count when simulating these inputs is significantly smaller than the reference inputs. They propose three reduced inputs: *smred* for short simulations (100 million instructions), *mdred* for medium length simulations (500 million instructions), and *lgred* for full-length, reportable simulations (1 billion instructions). For determining whether two input sets result in more or less the same behavior, KleinOsowski and Lilja used the chi-squared statistic based on the function-level execution profiles for each input set. A function-level profile is nothing more than a distribution that measures what fraction of the time is spent in each function during the execution of the program. Measuring these function-level execution profiles can for example be done using the UNIX utility gprof. A resemblance of

---

* http://www-mount.ee.umn.edu/~lilja/spec2000

function-level execution profiles does not necessarily imply a resemblance of other workload characteristics that are probably more closely related to performance, such as instruction mix, cache behavior, branch predictability, and similar characteristics. For example, consider the case when we would scale down the total time spent in each function by a factor *S*. Obviously, this will result in the same function-level execution profile because the function-level execution profile quantifies the time spent in each function *relative* to the total execution time. However, a similar function-level execution profile does not guarantee a similar data stream behavior. For example, reducing the input set might reduce the data memory footprint potentially leading to a significantly different data cache behavior. KleinOsowski and Lilja also recognized that this is a potential problem.

The workload analysis methodology presented in this chapter can also be used for validating the representativeness of reduced inputs, see also Eeckhout et al. [5] for more details. Indeed, if a reduced input is situated close to the reference input in the rescaled PCA space, we conclude that the reduced input results in similar behavior. The dendrogram obtained from the workload analysis of MinneSPEC is shown in Figure 9.9. For example,

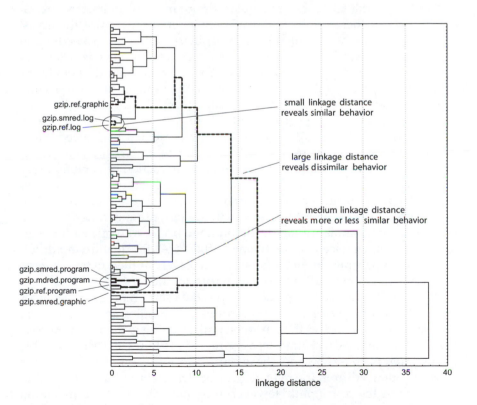

Figure 9.9 Cluster analysis for validating reduced inputs.

although the *smred* input for gzip.log has a dynamic instruction count that is more than a factor 50 smaller than the reference input and a data memory footprint that is also a factor 50 smaller, *smred* results in a similar program behavior as the *ref* input. As such, using the *smred* input instead of the reference input will result a simulation speedup of a factor 50 without loosing accuracy. The *smred* input for gzip.graphic on the other hand, yields a program behavior that is dissimilar to the reference input and should therefore not be considered as a viable alternative for the reference input.

From this MinneSPEC validation study, we can make some general conclusions. The *lgred* input is generally the best input among the inputs proposed in MinneSPEC. In other words, *lgred* generally yields (more or less) similar behavior. Unfortunately, there are a few exceptions, namely mcf, gcc and vortex. The smallest inputs, *smred* and *mdred*, generally lead to dissimilar behavior except for gzip.source, gzip.random and gzip.log.

9.5   Related work

The methodology presented in this chapter is built on the idea of measuring benchmark (dis)similarity. Indeed, program-input pairs that are close to each other in the workload space are similar. Program-input pairs that are far away from each other exhibit dissimilar behavior. In the literature, there exist only a few approaches to measuring benchmark similarity. Saveedra and Smith [12] present a metric that is based on dynamic program characteristics for the Fortran language. In their metric, Saveedra and Smith include the instruction mix, the number of function calls, the number of address computations, other such characteristics. For measuring the difference between benchmarks they used the squared Euclidean distance. The methodology presented in this chapter differs from the one presented by Saveedra and Smith for at least two reasons. First, the program characteristics used in our methodology are more suited for performance prediction of contemporary architectures because we include branch prediction accuracy, cache miss rates, ILP, and so forth—these characteristics are more closely related to performance; we thus expect that the methodology presented here yields a better benchmark similarity metric. Second, we prefer to work with uncorrelated program characteristics (obtained after PCA) for quantifying differences between program-input pairs. As is extensively argued in this chapter, removing correlation yields a better similarity metric.

Hsu et al. [6] studied the impact of input data sets on program behavior using high-level metrics, such as procedure-level profiles and IPC, as well as low-level metrics, such as the execution paths leading to data cache misses. They conclude that the test input set as provided in SPEC CPU is not suitable for simulation purposes because the execution profile is quite different from the profile obtained from the reference input. The train input was found to be better than the test input. However, they observed that the execution paths leading to data cache misses are very different between the train input and the reference input.

Yi and Lilja [17] propose a technique for classifying benchmarks with similar behavior, that is, by grouping benchmarks that stress the same processor components to similar degrees. Their method is based on a Plackett-Burman design. A Plackett-Burman design is a technique that allows researchers to measure the impact of variables by making a limited number of measurements. For example, consider the case where we want to measure the impact of n variables where each variable can have b unique values. The total number of experiments (or in our case simulations) that need to be done is b^n. This is also called a full multifactorial design. A Plackett-Burman design on the other hand is a fractional multifactorial design. It is a well-established technique for measuring the impact of n variables and their interactions by doing a limited number of experiments, namely $2(n+1)$. This is done by varying all parameters simultaneously in a well-chosen foldover design. For a more elaborate discussion of this type of analysis, see Chapter 5.

The issues discussed in this chapter are also related to sampling and simulation points (see also Chapters 6 and 7). Both sampling and simulation points need to find a number of samples or simulation points so that they are representative of the complete benchmark execution using the reference input. Several proposals have been made to quantify the representativeness of sampled execution and simulation points. Iyengar et al. [8] propose an R-metric for measuring the representativeness of a sampled trace. Lafage and Seznec [11] choose representative samples using cluster analysis. They applied their method to data cache simulations. Characterizing the individual samples is done using two microarchitecture-independent metrics, one that captures the temporal locality of the memory reference stream and one that captures the spatial locality of the memory reference stream. Sherwood et al. [13] characterize the large-scale behavior (as seen over billions of instructions) of computer programs using one microarchitecture-independent metric, namely the *basic block vector (BBV)*. In essence, the BBV quantifies the basic block execution profile. By measuring a BBV for each program slice (containing, for example, 100 million instructions), the various program slices can be characterized. Subsequently, the program slices with similar BBVs and thus similar behavior are grouped together through clustering. For each cluster, a representative sample can be chosen. The latter approach is discussed in more detail in Chapter 7.

Another possible application of data reduction techniques such as principal components analysis is to compare different workloads. Chow et al. [1] used PCA to compare the branch behavior of Java and non-Java workloads. The interesting aspect of using PCA in this context is that PCA is able to identify why two workloads differ. This can be done by analyzing the principal components. Chow et al. conclude, for example, that Java workloads tend to have more indirect calls whereas non-Java workloads tend to have more direct and indirect jumps.

Huang and Shen [7] quantify the impact of input data sets on the bandwidth spectrum of computer programs. The bandwidth spectrum measures the average bandwidth requirements of a program's instruction and data stream as a function of the available local memory. They conclude that the basic shape of the bandwidth spectrum does not change much with varying inputs.

9.6 Summary

This chapter presented a workload analysis methodology to measure benchmark similarity. It is based on two multivariate data analysis techniques, principal components analysis (PCA), and cluster analysis (CA). In the first step, a number of workload characteristics need to be measured, after which principal components analysis is performed to remove correlation from the data set. By retaining the most significant principal components, we achieve an important data reduction. Typically, the workload space is transformed from a 30-dimensional space into a 2- to 6-dimensional space. This workload space view gives us an opportunity to study benchmark similarity. Benchmarks that are close to each other suggest similar behavior; benchmarks that are more far away from each other suggest dissimilar behavior. The next step in this methodology is to perform cluster analysis on the transformed workload space. CA yields a dendrogram that clearly shows the similarities between the various benchmark clusters in the workload space.

We have discussed three important applications for this workload analysis methodology. The first application is gaining insight in program behavior. We discussed two examples, (1) studying the impact of input data sets on program behavior and (2) studying the interaction between Java applications and virtual machines (VMs). For the first example, strong clustering suggests a small impact of the input on overall behavior, whereas weak clustering suggests a large impact on overall behavior. For the second example, strong clustering for a single VM and multiple benchmarks suggests that overall behavior is primarily determined by the VM. This is typically observed for short-running Java applications. For long-running Java applications, weak clustering is observed, which suggests that the overall behavior then is primarily determined by the application.

Our second application is workload design or the composition of a set of representative benchmarks that covers the complete workload space. The basic idea is to select a representative benchmark per cluster as determined through CA. As such, only one benchmark is picked per cluster of similarly behaving benchmarks. We have shown an example in which we reduced the workload from 79 benchmark-input pairs to 16 benchmark-input pairs resulting in a simulation speedup of a factor 7.8.

Our third application is the validation of reduced input data sets. Previous work proposed to modify the input to a benchmark in order to reduce the dynamic instruction count of the benchmark. This is done by reducing the number of iterations of the algorithm, reducing the data set, truncating the input, etc. The workload analysis methodology discussed in this chapter can be used to measure the similarity between the reduced input set and the reference input set.

References

1. Chow, K., Wright, A., and Lai, K., Characterization of Java workloads by principal components analysis and indirect branches, *Proceedings of the Workshop on Workload Characterization (WWC) held in conjunction with the 21st Annual ACM/IEEE International Symposium on Microarchitecture (MICRO-31)*, 11–19, Nov. 1998.
2. Eeckhout, L., Georges, A., and De Bosschere, K., How Java programs interact with virtual machines at the microarchitectural level, *Proceedings of the 18th International Conference on Object-Oriented Programming Systems, Languages and Applications (OOPSLA)*, 169–186, Oct. 2003.
3. Eeckhout, L., Vandierendonck, H., and De Bosschere, K., Workload design: Selecting representative program-input pairs, *Proceedings of the 2002 International Conference on Parallel Architectures and Compilation Techniques (PACT 2002)*, 83–94, Sept. 2002.
4. Eeckhout, L., Vandierendonck, H., and De Bosschere, K., Quantifying the impact of input data sets on program behavior and its applications, *Journal of instruction-level parallelism*, http://www.jilp.org/vol5, 5(Feb.), 2003.
5. Eeckhout, L., Vandierendonck, H., and De Bosschere, K., Designing computer architecture research workloads, *IEEE Computer*, 36, 2, 65–71, 2003.
6. Hsu, W.C., Chen, H., Yew, P.Y., and Chen, D.-Y., On the predictability of program behavior using different input data sets, *Proceedings of 6th Workshop on Interaction between Compilers and Computer Architectures (INTERACT 2002)*, held in conjunction with the *8th International Symposium on High-Performance Computer Architecture (HPCA-8)*, 45–53, Feb. 2002.
7. Huang, A.S. and Shen, J.P., The intrinsic bandwidth requirements of ordinary programs, *Proceedings of the Seventh International Conference on Architectural Support for Programming Languages and Operating Systems (ASPLOS-VII)*, 105–114, Oct. 1996.
8. Iyengar, V.S., Trevillyan, L.H., and Bose, P., Representative traces for processor models with infinite cache, *Proceedings of the Second International Symposium on High-Performance Computer Architecture (HPCA-2)*, 62–73, Feb. 1996.
9. Johnson, R.A. and Wichern, D.H., *Applied Multivariate Statistical Analysis*, Prentice Hall, Englewood Cliffs, NJ, 5th edition, 2002.
10. KleinOsowski, A.J. and Lilja, D.J., MinneSPEC: A new SPEC benchmark workload for simulation-based computer architecture research, *Computer Architecture Letters*, 10–13, June 2002.
11. Lafage, T. and Seznec, A., Choosing representative slices of program execution for microarchitecture simulations: A preliminary application to the data stream, *IEEE 3rd Annual Workshop on Workload Characterization (WWC)*, held in conjunction with the *International Conference on Computer Design (ICCD)*, Sept. 2000.
12. Saavedra, R.H. and Smith, A.J., Analysis of benchmark characteristics and benchmark performance prediction, ACM Transactions on Computer Systems, 14, 4, 344–384, 1996.
13. Sherwood, T., Perelman, E., Hamerly, G., and Calder, B., Automatically characterizing large scale program behavior, *Proceedings of the 10th International Conference on Architectural Support for Programming Languages and Operating Systems (ASPLOS-X)*, 45–57, Oct. 2002.

14. Shuf, Y., Serrano, M.J., Gupta, M., and Singh, J.P., Characterizing the memory behavior of Java workloads: A structured view and opportunities for optimizations, *Proceedings of the 2001 ACM SIGMETRICS International Conference on Measurement and Modeling of Computer Systems*, 194–205, June 2001.
15. Srivastava, A. and Eustace, A., ATOM: A system for building customized program analysis tools, *Technical Report 94/2*, Western Research Lab, Compaq, March 1994.
16. StatSoft, Inc. STATISTICA for windows: Computer program manual, http://www.statsoft.com, 1999.
17. Yi, J., Lilja, D.L., and Hawkins, D.M., A statistically rigorous approach for improving simulation methodology, *Proceedings of the Ninth International Symposium on High Performance Computer Architecture (HPCA-8)*, 281–291, Feb. 2003.

Chapter Ten

Introduction to Analytical Models

Eun Jung Kim, Ki Hwan Yum, and Chita R. Das

Contents

10.1 Introduction

Performance analysis of a computer system is required to provide intelligent answers to a variety of questions that arise during various stages of its life cycle. The three techniques used for performance analysis are measurement, simulation, and analytical modeling. Measurement is the most credible approach to accurately evaluate the performance of a system, but it is very costly and many times not feasible simply because the target system may not be available or not yet built. Thus, simulation is an effective technique to predict the performance of any existing or new system. A simulation program, usually written in a high-level programming language, still requires tremendous amount of time and computing resources to model a relatively large or complex system precisely. On the other hand, analytical models attempt to capture the behavior of a computer system quite effectively and can provide quick answers to many questions. However, the model can become intractable as the system complexity increases.

For example, a very simple model of the average time required for a processor to access the main memory in the event of a cache miss is given by

$$T = T_c + (1 - h) \times T_m.$$

In this model, T is the average memory access time that we are trying to determine, h is the fraction of all memory references that hit in the cache (the hit ratio), T_c is the time required to read a block from the cache, and T_m is the time required to read a block from the main memory. Although this model is an over-simplification of how the memory system of a computer behaves, it does provide some insights into the effect that certain parameter changes, such as that the cache hit ratio or the memory access time will have on the average memory access time.

Although simple analytical models provide a bird's-eye view of a target system, sophisticated analytical models can be developed to obtain detailed information about how the system may behave under different conditions. For example, such models can be used to find an upper bound on the number of requests that a system is capable of handling or to quantify the average queueing time of jobs inside the system. These types of models are also useful for quickly analyzing the impact of system and workload parameters on system performance.

The motivation of this chapter is to give a quick review of the mathematical techniques that can be used for performance analysis of computer systems. It contains two different techniques that have been widely used: queueing theory and Petri net model. *Queueing theory* is the study of waiting in line [1], and is used to predict, for example, how long a job stays in a queue, how many jobs are in the system, and what is the system throughput. Petri net [2], on the other hand, is a graphical and mathematical modeling tool, which is particularly useful for capturing concurrency and synchronization behaviors [3].

10.2 Queueing theory

This section provides a concise review of the important queueing models used in analyzing computer systems. First, we summarize the relevant stochastic process concept that is required to understand queueing systems. Then, we describe a generic queueing system and introduce some fundamental laws of queueing theory before summarizing the main results of single queue systems. Finally, the solution techniques for network of queues are included to aid in analyzing complete systems.

10.2.1 Stochastic processes

In studying queueing systems, we need to deal with not only several random variables but also several different sequences that are functions of time. For example, the number of jobs in a queue at time t, $n(t)$, is a random variable. To specify its behavior, we need to know the probability distribution function for $n(t)$. Similarly, the waiting time in a queue, $w(t)$, is a random time dependent parameter. Such random functions of time, or sequences are called *stochastic processes*. Such processes are used to present the behavior of a queueing system in terms of its state space. The types of stochastic processes that are used in queueing system analysis are the following:

- Discrete-State and Continuous-State Processes: Depending on the values that the random variables can take, the process can be discrete or continuous. For example, $n(t)$ is a discrete-state process, whereas $w(t)$ is a continuous-state process.
- Markov Processes: If the future states of a process are independent of its past and depends only on the present state, the process is called a Markov process. A discrete-state continuous-time Markov process is called a Markov chain. The most important property of a Markov process is its memoryless behavior, which is captured by the exponential distribution for continuous-time processes and geometric distribution for discrete-time processes.
- Birth-Death Processes: A Markov process in which the transitions are restricted to neighboring states only is called a birth-death process.
- Poisson Processes: If the interarrival times of a Markov process are independent and identically distributed (IID), the number of arrivals n over a given time interval $(t, t + x)$ has a Poisson distribution. The arrival process is called a Poisson process, or Poisson stream. A Poisson process has the following properties:
  (a) Merging of k Poisson processes with mean rates $\lambda_i (i = 1, 2, ..., k)$ results in a Poisson process with a mean rate given by

$$\lambda = \sum_{i=1}^{k} \lambda_i$$

(b) If a Poisson process with a mean rate λ is split into k subprocesses such that the probability of a job going to the ith subprocess is p_i, each subprocess is also a Poisson process with a mean rate of $p_i\lambda$.

(c) If the arrivals to a single server with exponential service time are a Poisson process with a mean rate, the departures are also a Poisson process with the same rate, provided that the arrival rate is less than the service rate μ.

(d) If the arrivals to a service center with m servers are Poisson with a mean rate, the departures also constitute a Poisson stream with the same rate, provided that the arrival rate is less than the total service rate $\sum_i \mu_i$ of all m servers.

With these definitions, let us now look at a generic queueing system.

10.2.2 A generic queueing system

A generic queueing system is represented by a six-tuple notation, given by A/S/m/B/N/SD, where the first term stands for the arrival process, the second term represents the service time distribution, the third term denotes number of servers, the fourth term represents the buffer or queue size, the fifth term represents the population size, and the last term represents the service discipline [4]. A general queueing system depicting the six terms is shown in Figure 10.1.

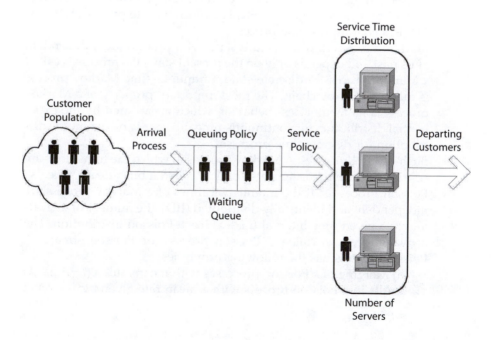

Figure 10.1 A Generic queueing system.

The arrival process to a queueing system is characterized by the inter-arrival time distribution of the customers to the system. The interarrival times are usually assumed to be IID random variables. Similarly, the amount of time a customer spends at a server is given by the underlying service time distribution. Although any known distribution can be used to represent the arrival and service time distributions, the commonly used distributions are exponential or memoryless (M), deterministic (D), and general (G). Other distributions such as Erlang and hyperexponential have been used to capture the service time variation of computer systems [5].

The third parameter, number of servers, specifies how many concurrent jobs can be served by the service center, and the fourth parameter captures the actual buffering capacity of the server. The fourth parameter is usually dropped from a queueing system notation if the buffer size is sufficiently large so that it can be considered an infinite capacity. Similarly, if the customer population is large, it can be assumed as infinite, and the fifth parameter is dropped from the system representation. The last term, service discipline (SD), represents the order in which customers are serviced at the center. Starting from the most common first-come-first-served (FCFS), scheduling to other prioritized policies, such as round robin (RR) and process sharing (PS), can be captured by this parameter.

Based on this notation, a few queueing systems are described next.

M/M/1: This is the simplest queueing system to analyze. The arrival and service times are exponentially distributed (Poisson processes), and the system consists of only one server. This queueing system can be applied to a wide variety of problems because any system with a very large number of independent customers can be approximated as a Poisson process. However, exponential service time distribution is not realistic for many applications and, thus, is only a crude approximation.

M/D/n: The arrival process is a Poisson process and the service time distribution is deterministic. The system has n servers (e.g., a ticket booking counter with n cashiers), and the service time is the same for all customers.

G/G/n: This is the most general queueing system, where the arrival and service time distributions are both arbitrary. The system has n servers. This is the most complex system, for which no analytical solution is known.

For such queueing systems, we are interested in determining output parameters such as the average number of customers in the system (or in the queue), average response time, and throughput of the system. Although the arrival and service time distributions are required to analyze a queueing system as a stochastic process for in-depth understanding of the system behavior, many times a set of simple relationships can be used for quick estimate of the system parameters. These relationships, called operational laws [6,7] or fundamental laws, are summarized in the following subsection.

10.2.3   Fundamental laws

A large number of day-to-day problems in computer system performance analysis can be solved by using some simple relationships that do not require any assumptions about the distribution of service times and arrival times. Several such relationships, called operational laws, were identified originally by Buzen (1976) [6] and later extended by Denning and Buzen (1978) [7]. *Operational* means directly measurable quantities such as throughput and arrival rate.

Let us assume that A is the number of arrivals during time T to the queueing system, depicted in Figure 10.1, C is the number of completion during this observation period, and B is the system busy time. Using these measured quantities, we can define the following simple relations:

- Arrival rate

$$\lambda = \frac{A}{T}$$

- Throughput

$$X = \frac{C}{T}$$

- Utilization

$$U = \frac{B}{T}$$

- Mean service time

$$S = \frac{B}{C}$$

- Utilization law

$$U = \frac{B}{T} = \frac{C}{T} \times \frac{B}{C} = XS$$

10.2.3.1   Forced flow law

In a queueing system with several resources, the number of arrivals to a device i is equal to its number of departures,

$$A_i = C_i.$$

Simplification of this yields the throughput of device i as

$$X_i = \frac{C_i}{T} = XV_i,$$

where V_i denotes the number of visits (visit count) to device i. This is called the forced flow law because it enforces a balance system.

10.2.3.2   Little's law

Little's law states that the average number of customers (N) can be determined as

$$N = \lambda R, \tag{1}$$

where is the average customer arrival rate and R is the average service time of a customer. The proof of this theorem can be found in any standard textbook on queueing theory [1]. Here, we will focus on an intuitive understanding of the result. Consider an example of a restaurant where the customer arrival rate (λ) doubles, but the customers still spend the same amount of time in the restaurant (R). This will double the number of customers in the restaurant (N). By the same logic, if the customer arrival rate remains the same, but the customers service time doubles, this will also double the total number of customers in the restaurant. Little's law is an important relation that relates the three important parameters of a queueing system.

10.2.3.3   General response time law

Using Little's law, we have

$$Q = XR,$$

where Q is the total number of customers in the system and the job flow is balanced so that arrival is equal to departure. If Q_i denotes the queue length at device i, we have $Q = Q_1 + Q_2 + \cdots + Q_M$ for a system with M queues.

Because Little's law can be applied to any device as long as the balance of the flow is maintained, we have

$$XR = X_1 R_1 + X_2 R_2 + \cdots + X_M R_M.$$

Dividing both sides by X and using forced flow law, we get the average response time R as

$$R = \sum_{i=1}^{M} R_i V_i.$$

This is called the general response time law, where the system response time is the summation of the product of the response time and visiting counts at each of the device. This is a very intuitive yet extremely important result that relates the three system parameters.

10.2.3.4   Interactive response time law

In an interactive system, where users have an average think time Z before generating requests that are serviced by the system with an average response time R, the total time spent in the system is $R + Z$. Each user generates about $T/(R + Z)$ requests in the time period T. $(R + Z)$ is the average response time per request, and thus, $1/(R + Z)$ is the number of requests per unit time. System throughput is expressed as

$$X = \frac{N[T/(R+Z)]}{T} = \frac{N}{R+Z}$$

or

$$R = (N/X)\ Z.$$

10.2.4   The M/M/1 queue

In this subsection we summarize the $M/M/1$ queueing system. As we have seen earlier, $M/M/1$ refers to negative exponential arrival and service times with a single server. This is the most widely used queueing system and is a good approximation for a large number of systems. It assume a Poisson arrival process, which is a very good approximation for the arrival processes in real systems that meet the following conditions:

- The number of customers in the system is very large;
- Impact of a single customer on the performance of the system is very small, that is, a single customer consumes a very small percentage of the system resources;
- All customers are independent, that is, their decisions to use the system are independent of each other.

Now that we have established scenarios where we can assume an arrival process to be Poisson, let us look at the probability density distribution for a Poisson process. The probability of seeing n arrivals in a period 0 to t is given by

$$P_n(t) = \frac{(\lambda t)^n}{n!}\, e^{-\lambda t},$$

where t is the time interval 0 to t and n is the total number of arrivals in the interval 0 to t, and is the average arrival rate in arrivals per sec.

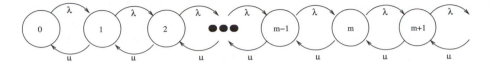

Figure 10.2 State transition diagram of the *M/M/1* system.

Suitability of an *M/M/1* queue is easy to identify from the server stand-point. For example, a single transmitting queue feeding an outgoing link is a single server and can be modeled as an *M/M/1* queueing system. If a single transmit queue feeds two load-sharing links, it should be modeled as an *M/M/2* queue.

Queueing models are usually solved using Markov Chain (MC) models. For example, an *M/M/1* queue is represented by the following state transition diagram in Figure 10.2, where each state of the queue gives the number of jobs in the system. Arrival of a job, with an average rate, increases the number of jobs in the system, whereas departure of a job after completing service at the server reduces the number of jobs by one. The solution of this MC gives the probabilities of all states in the system [5,4]. The state probabilities are used to estimate the performance parameters.

First we define the traffic intensity (sometimes called occupancy) as $\rho = (\lambda/\mu)$. For a stable system, the average service rate should always be higher than the average arrival rate. (Otherwise the queues would grow indefinitely). Thus, ρ should always be less than one. Also, note that because we are using average rates here, the instantaneous arrival rate may exceed the service rate. Over a longer time period, the service rate should always exceed the arrival rate.

The solution of the above MC gives the state probabilities, which can be used to find the performance parameters.

Now, using the state probabilities, the mean number of customers in the system (N) becomes

$$E[N] = \sum_{i=1}^{\infty} np_n = \frac{\rho}{1-\rho}(1).$$

Note that as approaches 1, the number of customers would become very large. This can be easily justified intuitively. The value will approach 1 when the average arrival rate starts approaching the average service rate. In this situation, the server would always be busy and a lead to a queue build up.

The average response time including service time, is computed using Little's Law as $N = R$ or

$$E[R] = \frac{1}{\mu - \lambda}.$$

Again we see that as mean arrival rate (λ) approaches mean service rate (μ), the waiting time becomes very large. Figure 10.3 depicts the response

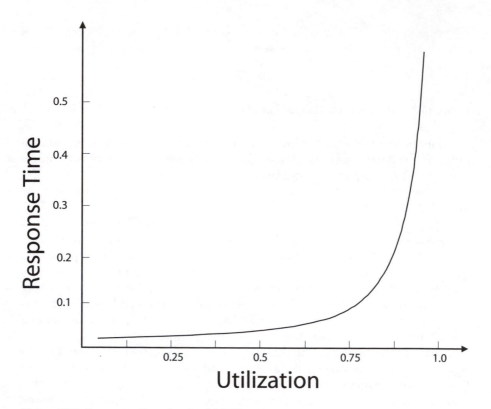

Figure 10.3 Response time for the *M/M/1* system.

time behavior of an *M/M/1* queue. For other parameters, the readers should refer to [5,4].

10.2.5   The M/M/m queue

The *M/M/m* queueing system is identical to the *M/M/1* system except that there are *m* servers. A customer at the head of the queue is routed to any server that is available. The Markov model for the *M/M/m* system is given later.

As shown in Figure 10.4, in the first *m* stages, the service rate increases linearly because more servers are available. After the *m* customers are busy,

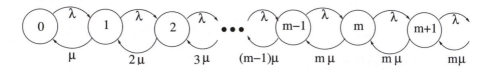

Figure 10.4 State transition diagram of the *M/M/m* system.

the service rate remains constant at $m\mu$. Using the state diagram we obtain the state probabilities as

$$
p_n = \begin{cases}
p_0 \dfrac{(m\rho)^n}{n!}, & n \leq m \\[3mm]
p_0 \dfrac{m^m \rho^n}{m!}, & n > m,
\end{cases}
$$

where ρ is given by

$$
\rho = \frac{\lambda}{m\mu} < 1.
$$

We can calculate p_0 using the relation of $\sum_{n=0}^{\infty} p_n = 1$, which gives

$$
p_0 = 1 \bigg/ \left(1 + \frac{(m\rho)^m}{m!(1-\rho)} + \sum_{n=0}^{m-1} \frac{(m\rho)^n}{n!} \right).
$$

The probability that a customer has to wait is denoted by

$$
P_Q = \sum_{n=m}^{\infty} p_n = \sum_{n=m}^{\infty} \frac{p_0 m^m \rho^n}{m!}
$$

$$
= \frac{p_0 (m\rho)^m}{m!} \sum_{n=m}^{\infty} \rho^{n-m}
$$

and reduces to

$$
P_Q \approx \frac{p_0 (m\rho)^m}{m!(1-\rho)}.
$$

This is known as the well-known *Erlang C formula*. This equation is widely used in telephony to estimate the probability of a call request finding all of the m lines busy.

We can now calculate the mean number of customers (N) in the system. It is easier to calculate the number of customers in the queue and number in service separately. The number of customers in the queue is

$$
E[N] = \sum_{n=m+1}^{\infty} (n-m) p_n = P_Q \frac{\rho}{1-\rho}
$$

and the number of customers in the server is

$$E[N] = \sum_{n=0}^{m-1} np_n + \sum_{n=m}^{\infty} mp_n = m\rho.$$

The total number of customers in the system is

$$E[N] = E[n_q] + E[n_s] = m\rho + P_Q \frac{\rho}{1-\rho}.$$

The service time is computed as

$$E[R] = \frac{E[N]}{\lambda} = \frac{1}{\mu}\left(1 + \frac{P_Q}{m(1-\rho)}\right).$$

10.2.6 *M/M/m/B Queue with finite buffer*

The $M/M/m/B$ system is similar to the $M/M/m$ queue except that the number of buffers B is finite. After the B buffers are full, all arrivals are lost. We assume that B is greater than or equal to m.

The state transition diagram for an $M/M/m/B$ queue is shown in Figure 10.5. The arrival and service rates in the system with n jobs are

$$\lambda_n = \lambda, \quad n = 0,1,2,\dots,B-1$$

$$\mu_n = \begin{cases} n\mu, & n = 0,1,2,\dots,m-1, \text{and} \\ \\ m\mu, & n = m, m+1,\dots,B. \end{cases}$$

The probability of n jobs in the system in terms of the traffic intensity $\rho = \lambda/m\mu$ is

$$p_n = \begin{cases} \dfrac{(m\rho)^n}{n!} p_0, & n = 1,2,\dots,m-1, \text{and} \\ \\ \dfrac{\rho^n m^m}{m!} p_0, & n = m, m+1,\dots,B. \end{cases}$$

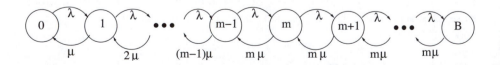

Figure 10.5 Discrete-time Markov chain for the $M/M/m/B$ system.

The probability of zero jobs in the system is calculated by the following equation.

$$\sum_{n=0}^{B} p_n = 1.$$

This gives

$$p_0 = \left[1 + \frac{1 - \rho^{B-m+1}(m\rho)^m}{m!(1-\rho)} + \sum_{n=1}^{m-1} \frac{(m\rho)^n}{n!} \right]^{-1}.$$

The mean number of customers (N) in the system is

$$E[N] = \sum_{n=1}^{B} n p_n.$$

For $m = 1$,

$$E[N] = \frac{\rho}{1-\rho} - \frac{(B+1)\rho^{B+1}}{1-\rho^{B+1}}.$$

The mean number of jobs in the queue is

$$N_Q = \sum_{n=m+1}^{B} (n-m) p_n.$$

For $m = 1$,

$$N_Q = \frac{\rho}{1-\rho} - \rho \frac{1 + B\rho^B}{1-\rho^{B+1}}.$$

In the state $n = B$, the effective arrival rate is

$$\lambda' = \sum_{n=0}^{B-1} \lambda p_n = \lambda(1 - p_B).$$

The average service time becomes

$$E[R] = \frac{E[N]}{\lambda'} = \frac{E[N]}{\lambda(1 - p_B)}.$$

10.2.7   The M/G/1 queue

Many queueing models used in performance analysis assume exponential interarrival times and exponential service times, which are covered by $M/M/m$ systems in the previous sections. In a $M/G/1$ queue, general service times are used.

The parameters are

λ: arrival rate in jobs per unit time
$E[s]$: mean service time per job
C_s: coefficient of variation of the service time.

The traffic intensity, ρ, is defined as $E[s]$, and the system is stable if the traffic intensity is less than 1. The probability of zero jobs in the system is $p_0 = 1 - \rho$ and the mean number of jobs in the system is

$$E[N] = \rho + \rho^2(1 + C_s^2)/[2(1-\rho)].$$

This equation is known as the Pollaczek-Khinchin (P-K) mean-value formula. It should be noted that the mean number in the queue grows linearly with the variance of the service time distribution. The variance of number of jobs in the system is

$$\mathrm{Var}[n] = E[n] + \lambda^2 \mathrm{Var}[s] + \frac{\lambda^3 E[s^3]}{3(1-\rho)} + \frac{\lambda^4 (E[s^2])^2}{4(1-\rho)^2}.$$

Then the mean number of jobs in the queue is

$$E[n_q] = \rho^2(1 + C_s^2)/[2(1-\rho)]$$

and the variance of the number of jobs in the queue is

$$\mathrm{Var}[n_q] = \mathrm{Var}[n] - \rho + \rho^2.$$

The mean response time is

$$E[R] = E[N]/\lambda = E[s] + \rho E[s](1 + C_s^2)/[2(1-\rho)]$$

and the variance of the response time is

$$\mathrm{Var}[r] = \mathrm{Var}[s] + \lambda E[s^3]/[3(1-\rho)] + \lambda^2 (E[s^2])^2/[4(1-\rho)^2].$$

The mean waiting time is

$$E[w] = \rho E[s](1 + C_s^2)/[2(1-\rho)].$$

For example, if we consider the $M/D/1$ system then $C_s = 0$ and

$$E[w] = \rho E[s]/[2(1-\rho)].$$

10.2.8 Networks of queues

Although all the systems we have seen so far have only one queue, there exist many systems that contain multiple queues. Unlike single queueing systems, there is no easy notation for specifying the type of a queueing network. The simplest way to classify a queueing network is either that it is open or closed. An open queueing network has external arrivals and departures as shown in Figure 10.6. The jobs enter the system at "In" and exit at "Out." The number of jobs in the system varies with time. In analyzing an open system, we assume that the throughput is known and is equal to the total arrival rate, and the goal is to characterize the distribution of number of jobs in the system and average response time.

10.2.8.1 Solution of open queueing networks

With the assumption mentioned above, each queue can be analyzed independently. Open queueing networks are used to represent transaction processing systems, such as airline reservation systems or banking systems.

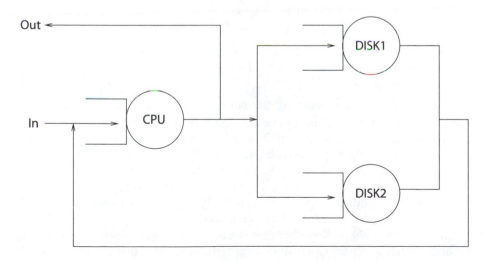

Figure 10.6 An open queueing network.

The key feature is that the arrival rate does not depend on the load of the system. For all fixed-capacity service centers in an open queueing network, the response time is

$$R_i = S_i(1 + Qi).$$

Intuitively, an arriving job sees Q_i jobs ahead of it, each of which will spend S_i time in service, thus the total response time is all of its predecessors plus itself. This is not an operational law because it assumes the memoryless property of the arrivals (i.e., Poisson arrival), which cannot be tested.

Assuming job flow balance, the throughput is equal to the arrival rate $X = \lambda$. The throughput of the ith device using the forced law is $X_i = XV_i$. The utilization of the ith device using the utilization law is

$$U_i = X_i S_i = XV_i S_i = \lambda D_i.$$

The queue length of the ith device using Little's law is

$$Q_i = X_i R_i = X_i S_i (1 + Q_i) = U_i (1 + Q_i)$$

or

$$Q_i = \frac{U_i}{1 - U_i}$$

and the response time for device i is

$$R_i = \frac{S_i}{1 - U_i}.$$

10.2.8.2   Closed queueing networks

On the other hand, a closed queueing network has no external arrivals or departures as shown in Figure 10.7. The jobs in the system keep circulating from one queue to the next. The total number of jobs in the system is constant. In analyzing a closed system, we assume that the number of jobs (N) is given, and we attempt to determine the throughput or the average service time. Two types of technique can be used to analyze closed network of queues: mean value analysis (MVA) and convolution*.

Mean value analysis allows solving closed queueing networks in a manner similar to that used for open queueing networks. Key differences are that

---

* For the convolution algorithm, please refer to the studies by Trivedi and Jain [5,4].

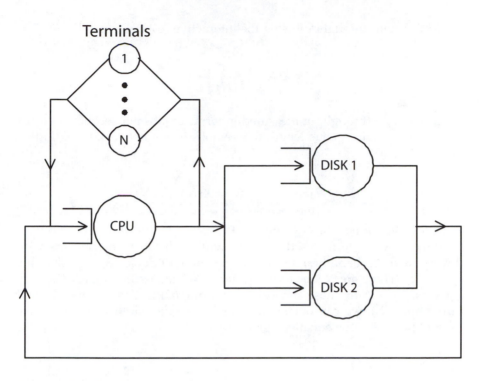

Terminals

DISK 1

CPU

DISK 2

Figure 10.7 A closed queueing network.

there are no outside arrivals and no departures, and the arrival rate at a device may depend on the load of the device. We will consider only fixed-capacity device here. Other variations are more complicated.

Given a closed queueing network with N jobs and the service is memoryless (exponential), Reiser and Lavenberg (1980) showed that the response time of the ith device is given by

$$R_i(N) = S_i[1 + Qi(N - 1)].$$

This form is similar to the open queueing network. The response time with N jobs depends on the one with $N - 1$ jobs. We can calculate the one with zero jobs in it and then extend to $1, 2, ..., N - 1, N$.

Given the response times at individual devices, the system response time using the general response time law is

$$R(N) = \sum_{i=1}^{M} V_i R_i(N).$$

The system throughput using the interactive response time law is

$$X(N) = \frac{N}{R(N) + Z}.$$

The device throughput measured in terms of jobs per second is $X_i(N) = X(N)V_i$. The device queue length with N jobs in the network using Little's law is

$$Q_i(N) = X_i(N)R_i(N) = X(N)V_iR_i(N).$$

Consider a model of a computer CPU connected to an input/output (I/O) device. We have a closed network with each job reentering the CPU directly with probability p_1 or after using the I/O device with probability $p_2 = 1 - p_1$. There are N jobs in the system. We can assume that $X_1(N) = \mu_1$, $X_2(N) = p_2\mu_1$. With these device throughput (CPU throughput X_1, I/O throughput X_2), we can obtain each device queue length and the system throughput using the equations given.

10.3 Petri nets

Petri nets are a graphical and mathematical modeling tool applicable to many systems [3]. A Petri net is a particular kind of directed graph, together with an initial state called the initial marking, M_0. The underlying graph Z of a Petri net is a directed, weighted, bipartite graph consisting of two kinds of nodes—called places and transitions—where arcs are either from a place to a transition or from a transition to a place. In graphical representation, places are drawn as circles, and transitions as bars or boxes. Arcs are labeled with their weights (positive integers), where a k-weighted arc can be interpreted as the set of k parallel arcs. Labels for unity weight are usually omitted. A marking (state) assigns to each place a nonnegative integer. If a marking assigns to place p a nonnegative integer k, we say that p is marked with k tokens. Pictorially, we place k black dots (tokens) in place p. A marking is denoted by M, an m-vector, where m is the total number of places. The pth components of M, denoted by $M(p)$, is the number of tokens in place p.

In modeling, places represent conditions and transitions represent events. A transition (an event) has a certain number of input and output places representing the preconditions and postconditions of the event, respectively. The presence of a token in a place is interpreted as holding the truth of the condition associated with the place. In another interpretation, k tokens are put in a place to indicate that k data items or resources are available.

A formal definition of a Petri net is as follows.

Definition 10.1

A Petri net is a five-tuple, PN = (P, T, F, W, M_0) where $P = \{p_1, p_2, ..., p_m\}$ is a finite set of places, $T = \{t_1, t_2, ..., t_n\}$ is a finite set of transitions, $F \subseteq (P \times T) \cup (T \times P)$ is a set of arcs (flow relation), $W:F \to \{1, 2, ...\}$ is a weight function, $M_0:P \to \{0, 1, 2, ...\}$ is the initial marking, $P \cap T = \varnothing$ and $P \cup T \neq \varnothing$.

A Petri net structure $Z = (P, T, F, W)$ without any specific initial marking is denoted by Z. A Petri net with the given initial marking is denoted by (Z, M_0).

In order to simulate the dynamic behavior of a system, a state or marking in a Petri nets is changed according to the following transition (firing) rule:

1. A transition t is said to be enabled if each input place p of t is marked with at least $w(p, t)$ tokens, where $w(p, t)$ is the weight of the arc from p to t.
2. An enabled transition may or may not fire (depending on whether or not the event actually takes place).
3. A firing of an enabled transition t removes $w(p, t)$ tokens from each input place p of t, and adds $w(t, p)$ tokens to each output place p of t, where $w(t, p)$ is the weight of the arc from t to p.

A transition without any input place is called a source transition, and one without any output place is called a sink transition. It should be noted that a source transition is unconditionally enabled and that the firing of a sink transition consumes tokens but does not produce any. A pair of a place p and a transition t is called a self-loop if p is both an input and output place of t. A Petri net is said to be pure if it has no self-loops. A Petri net is said to be ordinary if all of its arc weights are ones.

The firing of a transition may transform a Petri Net (PN) from one marking into another. With respect to a given initial marking M_0, the *reachability set* is defined as the set of all markings reachable through any possible firing sequences of transitions, starting from the initial marking [5]. The evolution of a PN can be completely described by its *reachability graph*, in which each marking in the reachability set is a node in the graph, while the arcs describe the possible marking-to-marking transitions. Arcs are labeled with the name of the transition whose firing caused the associated changes in the marking.

For the preceding rule of transition enabling, it is assumed that each place can accommodate an unlimited number of tokens. Such a Petri net is referred to as an infinite capacity net. For modeling many physical systems, it is natural to consider an upper limit to the number of tokens that each place can hold. Such a Petri net is referred to as a finite capacity net. For a finite capacity net (Z, M_0), each place p has an associated capacity $K(p)$, the maximum number of tokens that p can hold at any time. For finite capacity nets, for a transition t to be enabled, there is an additional condition that the number of tokens in each output place p of t cannot exceed its capacity $K(p)$ after firing$t$.

This rule with the capacity constraint is called the strict transition rule, whereas the rule without the capacity constraint is called the (weak) transition rule. Given a finite capacity net (Z, M_0), it is possible to apply either the strict transition rule to the given net (Z, M_0) or, equivalently, the weak transition rule to a transformed net $(Z, M_0$, the net obtained from (Z, M_0) by the following complementary-place transformation, where it is assumed that N is pure.

Step 1: Add a complementary place p' for each place p', where the initial marking of p' is given by $M_0'(p') = K(p') - M_0(p')$.

Step 2: Between each transition t and some complementary places p', draw new arcs (t, p') or (p', t) where $w(t, p') = w(p', t)$ and $w(p', t) = w(t, p)$, so that the sum of tokens in place p and its complementary place p' equals its capacity $K(p)$ for each place p, before and after firing the transition t.

10.3.1 Modeling with petri nets

Petri nets are used in the modeling of a specific class of problems, the class of discrete-event systems with concurrent or parallel events. Petri nets model systems, and particularly two aspects of systems—events and conditions—and the relationships among them.

For example, consider the following description of a computer system [8]:

- Jobs appear and are put on an input list. When the processor is free, and there is a job on the input list, the processor starts to process the job.
- When the job is complete, it is placed on an output list, and if there are more jobs on the input list, the processor continues with another job; otherwise it waits for another job.

We can identify several conditions of interest: The processor is idle; a job is on the input list; a job is being processed; a job is on the output list. Also we can identify several events: A new job enters the system; job processing is started; job processing is completed; a job leaves the system.

Figure 10.8 illustrates the modeling of this system. The "job enters" transition in this example is a source and the "job leaves" transition is a sink. This example shows several characteristics of Petri nets and the systems they can model. One is inherent *concurrency* or *parallelism*. There are two main kinds of independent entities in the system: the job and the processor. In the Petri net model, the events, which relate solely to one or the other, can occur independently; there is no need to synchronize the actions of the jobs and the processor. However, when synchronization is necessary (e.g., when both a job and an idle processor must be available for processing to start), the situation can be easily modeled.

Another major feature of Petri nets is their asynchronous nature. There is no inherent measure of time or the flow of time in a Petri net. This reflects a philosophy of time, which states that the only important property of

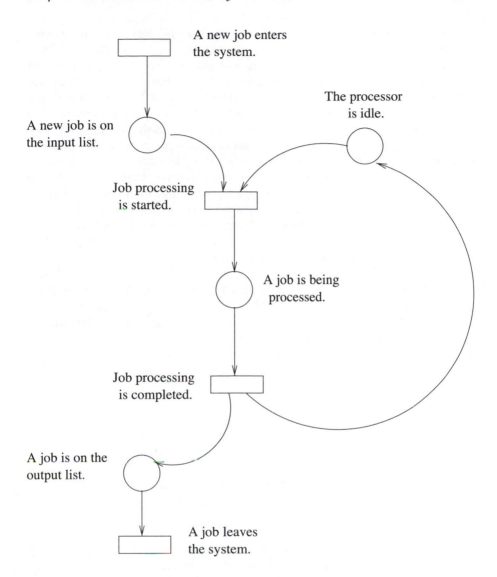

Figure 10.8 Modeling of a simple computer system.

time, from a logical point of view, is in defining a partial ordering of the occurrence of events. Events take variable amounts of time in real life; the Petri net model reflects this variability by not depending on a notion of time to control the sequence of events. Therefore, the Petri net structure itself must contain all necessary information to define the possible sequences of events of a modeled system. Thus in Figure 10.8, the event "Job processing is completed" must follow the corresponding event "Job processing is started" because of the structure of the Petri net, although no information is given or considered concerning the amount of time required to process a job.

A Petri net is viewed as a sequence of discrete events whose order of occurrence is one of possibly many allowed by the basic structure. If at any time more than one transition is enabled, then any of the several enabled transitions may fire. The choice as to which transition fires is made in a nondeterministic manner, that is, randomly or by forces that are not modeled. Although this nondeterminism is advantageous from a modeling point of view, it introduces considerable complexity into the analysis of Petri nets.

To reduce the complexity, one generally accepts that the firing of a transition (occurrence of an event) is considered instantaneous, that is, to take zero time. Because time is a continuous variable, the probability of any two or more events happening simultaneously is zero, and two transitions cannot fire simultaneously. The events modeled are considered primitive events. For example, in Figure 10.8 the event "Process a job" was modeled. But because this event is not a primitive one, it is decomposed into a *beginning* and an *ending*, which are instantaneous events, and the noninstantaneous occurrence.

The nondeterministic and nonsimultaneous firing of transitions in the modeling of concurrent systems takes two forms:

- Simultaneous events that may occur in any order (Figure 10.9(a));
- Conflicting transitions where the firing of one will disable the other (Figure 10.9(b)).

In Figure 10.9(a), the two enabled events do not affect each other in any way, and the possible sequences of events include some in which one event occurs first and some in which the other occurs first. In the other type of situation, shown in Figure 10.9(b), the two enabled transitions are in conflict. Only one transition can fire, because that removes the token from p and disables the other transition.

An important aspect of Petri nets is that they are uninterpreted models. The net of Figure 10.8 has been labeled with statements that indicate to a human observer the intent of the model, but these labels do not, in any way, affect the execution of the net. We only deal with the abstract properties inherent in the structure of the net.

Another valuable feature of Petri nets is their ability to model a system hierarchically. An entire net may be replaced by a single place or transition for modeling at a more abstract level (abstraction) or places and transitions may be replaced by subnets to provide more detailed modeling (refinement).

10.3.2   Stochastic petri nets

Stochastic Petri nets (SPNs) are obtained by associating stochastic and timing information to Petri nets [5]. We do this by attaching *firing time* to each transition, representing the time that must elapse from the instant that the

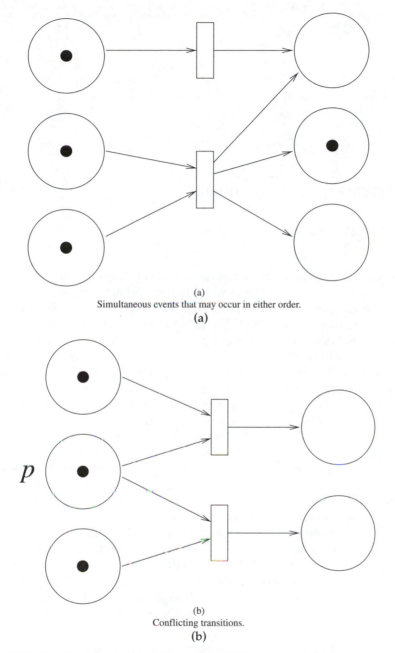

(a)
Simultaneous events that may occur in either order.
(a)

(b)
Conflicting transitions.
(b)

Figure 10.9 Simultaneous and conflicting transitions.

transition is enabled until the instant it actually fires in isolation, that is, assuming that it is not affected by the firing of other transitions. If two or more transitions are enabled at the same time, the firing of transitions is determined by the *race policy*; that is, the transition whose firing time elapses first is chosen

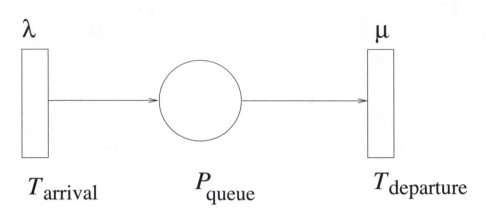

λ μ

T_{arrival} P_{queue} $T_{\text{departure}}$

Figure 10.10 SPN model of M/M/1 queue.

to fire next. If the firing times can have general distributions, SPNs can be used to represent a wide range of well-known stochastic processes. However, choices about execution policy and memory policy, besides the firing time distributions, must be specified. The firing times are often restricted to have an exponential distribution to avoid policy choices. A more important fact in this case is that an SPN can be automatically transformed into a continuous-time Markov Chain (CTMC). In a graphical representation, transitions with exponentially distributed firing times are drawn as rectangular boxes.

When an SPN is applied to performance analysis of computer networks, places can be used to denote the number of packets or cells in the buffer or the number of active users, or flows in the system, and the arrival and departure of packets, cells, users, or flows can be represented by firing or transitions.

In the following we show an SPN model for a simple $M/M/1$ queueing system to illustrate the transformation from an SPN into a CTMC.

In Figure 10.10, the number tokens in P_{queue} represents the number of customers in the system, including the one receiving service, if any. Whenever there is a customer (one or more tokens) in the system (in place P_{queue}), a customer may complete service when the transition T_{service} fires and the firing time is exponentially distributed with rate μ. The reachability graph of this SPN is, in fact, a simple birth-death process with an infinite state space, as shown in Figure 10.2. Measures such as system size, response time, and throughput can be computed by solving this CTMC.

10.4 Conclusions

In this chapter we have introduced queuing theory and Petri nets, which are important modeling tools for performance analysis of computer systems. Unlike other performance analysis techniques, such as measurement and simulation, analytical modeling can capture the behavior of a computer system effectively and can provide quick answers to many question. We have explained queuing theory and related fundamental laws to help the readers comprehend the basic principles. Then, according to queuing system classification, we discussed the parameters and related equations. The chapter also gives a quick review of Petri nets.

The motivation of this chapter is to introduce different analytic techniques that can be used in performance analysis of computer systems. It is impossible to provide an in-depth treatment of the vast area in a chapter. Interested readers should refer to the studies by Jain and Trivedi [4,5] for detailed analyses.

References

1. Kleinrock, L., *Queueing Systems Volume I: Theory.* Wiley-Interscience, Hoboken, NJ, 1975.
2. Petri, C.A., *Kommunikation mit Automaten.* Ph.D. thesis, Bonn: Institut für Instrumentelle Mathematik, Schriften des IIM Nr. 3, 1962. Also, English translation, Communication with automata, New York: Griffiss Air Force Base, Tech. Rep. RADC-TR-65-377, Vol. 1, Suppl. 1, 1966.
3. Murata, T., Petri nets: Properties, analysis and applications, *Proceedings of the IEEE 77*, 4, 541–580, 1989.
4. Jain, R., *The Art of Computer Systems Performance Analysis*, John Wiley & Sons, Inc., 1992.
5. Trivedi, K.S., *Probability and Statitics with Reliability, Queuing and Computer Science Applications.* Wiley-Interscience, Hoboken, NJ, 2002.
6. Buzen, J.P., A queueing network model of MVS, *ACM Computing Surveys 10*, 3, 319–331, 1978.
7. Denning, P.J., and Buzen, J.P., The operational analysis of queueing network models, *ACM Computing Surveys 10*, 3, 225–261, 1978.
8. Peterson, J.L., Petri nets *ACM Computing Survey 9*, 3, 223–252, 1977.

Chapter Eleven

Performance Monitoring Hardware and the Pentium 4 Processor

Brinkley Sprunt

Contents

11.1   Introduction

Performance-monitoring hardware is included in many of today's high-performance processors. This hardware can detect and count numerous performance events as the microprocessor executes. These event-count data provide a good understanding of how applications, the operating system, and the processor are performing and can be used to guide efforts to improve performance. This chapter describes the general capabilities of performance-monitoring hardware and provides examples of its use to obtain performance profiles and expose performance problems. This overview is followed with a description of the specific performance-monitoring features of the Intel Pentium 4 processor. The unique capabilities of the Pentium 4 performance-monitoring hardware and its improvements with respect to previous processors are discussed, including the event detectors and counters and support for obtaining nonspeculative event counts as well as precise, event-based profiles. Also, the performance-monitoring support for the simultaneous multithreaded execution features of the Pentium 4 Xeon processor are explained. The chapter then introduces the Brink and Abyss software tools for configuring the Pentium 4's performance monitoring hardware and collecting performance data. The unique approach taken by these tools to manage and document the complexity of the Pentium 4's performance monitoring features is also described. This chapter closes with a discussion of the issues facing the designers and users of performance monitoring hardware and software as microprocessor designs move toward more extensive use of chip multiprocessing and simultaneous multithreading designs.

11.2   Performance event monitoring

Performance-monitoring hardware (typically referred to as EMON, or *event-monitoring* hardware) provides a low-overhead mechanism for collecting processor performance data [15]. Once enabled, the EMON hardware can detect and nonintrusively count any of a set of performance events while the processor is running applications and the operating system. The available performance events can typically be grouped into five categories: program characterization, memory, pipeline stalls, branch prediction, and resource utilization. The program characterization events are used to characterize the attributes of a program (and/or the operating system) that are largely independent of the processor's implementation. The most common example of program characterization events are the number and type of instructions completed by the program (e.g., loads, stores, floating point, branches). The memory events often compose the largest event category and are used to analyze the performance of the processor's memory hierarchy. For example, memory events can be used to count references and misses to various caches and to count transactions on the processor-memory bus. The pipeline stall events are used to analyze how well the program's instructions flow through the pipeline. Processors with deep pipelines rely heavily on branch prediction hardware to keep the pipeline filled with useful instructions. The branch prediction events indicate how often the processor is able to predict accurately the outcome of branches and keep the correct instructions flowing into the processor's pipeline (e.g., by providing counts of mispredicted branches). The resource utilization events provide counts of the number of cycles a particular resource is in use or busy (e.g., the number of cycles a floating-point divider is being used).

11.2.1 EMON detectors and counters

EMON hardware is typically composed of two main components: event detectors and event counters. By properly configuring the event detectors and counters, a variety of events under various conditions can be obtained.

An event detector can be configured to detect any one of a large set of available events (e.g., cache misses or branch mispredictions). Often, event detectors have an event mask field that allows further qualification of the event to be specified. For example, the Pentium III event that detects load accesses to the level-2 cache (L2_LD) has an event mask that allows further qualification by the state of the cache line being accessed (i.e., modified, shared, exclusive, or invalid) [6]. The configuration of the event detector also allows qualification by the current privilege level of the processor. Operating systems use supervisor and user privilege levels to prevent applications from accessing and manipulating critical data structures and hardware that should only be directly used by the operating system. When the operating system is executing on the processor, it sets the privilege level to *supervisor*, and when the operating system selects an application to execute on the processor, it sets the privilege level to *user*. As such, the ability to qualify event detection by the processor's privilege level allows events to be detected that are caused only by the application or only by the operating system. By enabling event detection at either privilege level, events from both applications and the operating system will be detected.

In addition to counting the events detected by the performance event detectors, the performance event counters can also be configured to count only under certain edge and threshold conditions. The edge detection feature of the event counters is most often used for performance events that detect the presence or absence of certain conditions every cycle. For these events, an event count of one represents the presence of the condition and zero indicates the absence of the condition. For example, a pipeline stall event indicates the presence or absence of a pipeline stall every cycle. By counting the number of these events, one obtains the number of cycles the pipeline is stalled. However, it is often desirable to know how many stalls occurred (i.e., the number of times a stall began) rather than just the total number of cycles stalled. To enable the counting of stalls, the edge detect feature of the event counter can be used. When edge detect is enabled, the performance counter will increment by one only when the previous number of performance events reported by the event detector is less than the current number being reported by the event detector. In this manner, when the event detector reports zero events on a cycle followed by one event on the next cycle, the event counter has detected a *rising edge* and will increment by one. The sense of the edge detection can usually be inverted in order to count *falling edges* instead. For these events, disabling the edge detection feature counts stall durations and enabling edge detection counts stall occurrences. Additionally, by dividing the total stall duration count by the total stall occurrence count, one can obtain the average cycles stalled for a particular stalling condition.

The second major feature of the event counters is a generalization of the edge-detection support referred to as threshold support. With the threshold feature, the event counter compares a threshold value to the value being reported by the event counter on each cycle. If the value being reported by the event counter exceeds the threshold, then the counter will increment by one. The threshold feature is only of use for performance events that can report values greater than one on each cycle. For example, superscalar processors can complete more than one instruction per cycle. To determine how many times three or more instructions are completed per cycle, one could select the performance event for instructions completed and set the counter's threshold value to two and whenever three or more instructions complete in one cycle, the counter would increment by one.

11.2.2   Event-based sampling

Although performance event detectors and counters can easily be used to detect the presence of a performance problem and estimate its severity, to eliminate the performance problem, or to reduce its impact, it is necessary to determine the sections of code that are causing the performance problem. Once the programmer knows the code associated with the performance problem, the high-level algorithms used by the application and/or the low-level code to implement those algorithms can be altered to avoid performance problem entirely or lessen its severity.

EMON hardware can be used to create an event-based profile that identifies the code locations responsible for a particular performance problem. The approach used to create event-based profiles is similar to the approach used to create time-based profiles. A time-based profile indicates the code locations where an application spends the majority of its time and, by focusing performance-tuning efforts on these sections of the application, the benefits of performance tuning efforts are maximized. To create a time-based profile, a timer is configured to interrupt the application at regular intervals.

Upon each interrupt, the interrupt service routine (ISR) saves the value of the program counter. Once the application completes, the sampled program counter values are used to create a histogram of the number of samples versus code location. Assuming many program counter samples are collected, the histogram will show the most frequently executed sections of the program. The creation of an event-based profile is similar to a time-based profile, but instead of using a timer to interrupt the application at regular intervals of time, the performance monitoring hardware is configured to interrupt the application after a specific number of performance events have occurred. For example, one can configure performance monitoring hardware to interrupt the processor after every N occurrences of a particular performance event. The resulting event-based profile indicates the most frequently executed sections of code that caused the particular performance event.

To support event-based sampling, performance monitoring hardware typically provides the ability to cause a performance monitor interrupt upon the overflow of a performance event counter. In order to generate an interrupt after N occurrences of a performance event, the performance counter is initialized to a value of "overflow $-N$" before being enabled. A performance monitor ISR must also be installed to handle these interrupts. The job of the performance monitor ISR is to save away sample data from the program (e.g., the program counter) and to re-enable the performance event counter to cause another interrupt after N occurrences of the desired performance event. Upon completion of the application, the data samples saved by the performance monitor ISR can be used to create an event-based profile.

11.2.3   *Example uses of EMON hardware*

To illustrate the use and benefits of EMON hardware, let us examine two examples. The first example demonstrates how EMON data can be used to tune a high-level algorithm for better performance on a specific processor. The second example demonstrates how EMON data can be used to find a low-performing code sequence that can then be modified to improve its performance.

11.2.3.1   *Tuning high-level algorithms for different*
processor characteristics

Algorithms developed for one processor generation sometimes do not perform as well on a successive processor generation. Consider the following simple algorithm for finding prime numbers. The algorithm creates a large array where each element in the array represents an integer and begins by initializing all the array elements to zero. To find prime numbers, the algorithm repeatedly steps through the array with successively greater strides starting with a stride of two. On all but the initial steps, the array element is marked by storing a value of "1" into the element. The value of "1" indicates that the number corresponding to the array element is not prime. For example, the first pass through the array would use a stride of 2 and would mark locations 4, 6, 8, and so on, and the second pass would use a stride of three and mark locations 6, 9, 12, for example. Once all the passes through the array are complete, the array elements that are still zero correspond to prime numbers.

Now, consider how this algorithm performs on the original Pentium processor [8] and the Pentium Pro processor [7]. Both of these processors have an 8K L1 data cache, but the Pentium Pro processor has an additional 256K unified L2 cache. In general, one would expect the overall performance of the Pentium Pro to be better than Pentium because of its more advanced design and the presence of the large, unified L2 cache. However, a subtle difference in the allocation policies for these caches can cause some performance surprises. The Pentium processor does not allocate a new cache line for a store that misses the cache, but the Pentium Pro does. When a new line is allocated into the caches, the whole line (32 bytes of data) must be loaded from memory. In steady state, the prime number algorithm described earlier is accessing a large array (much larger than the caches) using a large stride. As such, almost every store to the array will miss the caches. On Pentium, each store miss results in one bus transaction between the processor and memory because a new cache line is not allocated. However, because the Pentium Pro allocates a new cache line on each store miss, each store miss causes four bus transactions to load the 32-byte cache line using the 8-byte-wide processor-memory bus. Also note that, in steady state, most of the lines in the Pentium Pro caches are dirty (they have data that is more recent than main memory) because they are holding the "1"s being stored into the prime number array. As such, when a new line is allocated, a dirty line must be written back to memory, requiring another four bus transactions. So, in steady state, the algorithm causes only one bus transaction for each store to the array for the Pentium processor, but causes eight bus transactions for the Pentium Pro processor. This 8x difference in required bus bandwidth is a bottleneck for the Pentium Pro processor and its performance on this algorithm is low compared to the Pentium processor.

Initially, the reasons for the lower performance of the Pentium Pro on the prime number algorithm were far from obvious. However, by collecting cache and bus performance data using the EMON hardware for these processors and examining the prime number algorithm, the problem described earlier soon became apparent. Once the problem was understood, it was noted that a slight change in the algorithm would substantially reduce the processor-memory traffic for the Pentium Pro processor. Instead of blindly storing a 1 into the array elements on each algorithm step, the array value should be tested and only if the value is 0 should a 1 be stored into the element. This simple, one-line change to the algorithm eliminates the majority of the dirty cache lines because each element in the prime number array will only be written once. This significantly reduces processor-memory traffic for the Pentium Pro by eliminating the write back of dirty cache lines to main memory and brings the Pentium Pro's performance to a level on par with or better than the Pentium. Without the performance-monitoring capabilities of these processors, the investigation and resolution of this performance problem would have been much more difficult.

11.2.3.2   Finding and eliminating poorly performing code sequences

As an example of how hardware EMON features can be used to identify the presence of a performance problem and locate the code that is causing the problem, consider the performance problem of partial register stalls on Intel's P6-based Pentium processors [6]. On these processors, a partial register stall occurs when the lower 8 or 16 bits of a 32-bit register are written by an instruction and then soon followed by another instruction that reads the full 32 bits from the register. On these processors, register renaming causes each write of a register to go to a different physical register. Consequently, the newly written 8 or 16 bits reside in a different physical register than the unchanged upper bits of the register. The full 32 bits of the register cannot be read until the partial components of the register's value are merged as the instructions retire, causing a stall of at least 7 cycles and typically 10-14 cycles.

To determine whether or not partial register stalls are present in an application, one can configure the EMON hardware to count the number of partial register stall cycles as the application runs. If the number of partial register stall cycles as a percentage of the total cycles for the application is significant (e.g., greater than 3%), this indicates that performance can be improved by finding and eliminating the partial register stalls in the application code.

Once partial register stalls are identified as a significant performance problem, one can use the event-based sampling (EBS) support of the performance counters to locate the most frequent occurrences of the partial register stalls. Once the samples have been collected, a partial-stall, event-based profile (essentially a histogram of sample counts versus code location) can be created. This profile enables the programmer to identify the specific locations in the application code that are causing the majority of the partial stalls. Once these locations are known, the programmer can modify the code to eliminate the partial register stalls.

11.3   Pentium 4 EMON

Intel's introduction of the Pentium 4 processor [4] marked a significant increase in the general capabilities found in microprocessor performance-monitoring hardware [14]. This section describes these key features of the Pentium 4 EMON hardware and describes its structure and interface.

11.3.1   Pentium 4 EMON features and improvements

The most noticeable advance in Pentium 4's EMON capabilities relative to prior processors is the inclusion of significantly more event detectors and counters. Before Pentium 4, microprocessors with EMON support typically included only two to six event detectors and counters. The Pentium 4 supports 48 event detectors and 18 event counters. This larger number of event counters enables much more performance data to be collected simultaneously than on previous processors, reducing the number of application runs to collect the desired data and reducing the inaccuracies introduced by multiple application runs.

The Pentium 4 also introduced instruction tagging mechanisms that allow nonspeculative performance event counts to be obtained. On speculative-execution processors, such as the Pentium 4, most instructions are executed before they are known to be on the correct execution path. Without a mechanism to separate the detected performance events into those caused by speculatively executed instructions that never complete from instructions that do complete (retire), erroneous conclusions about program behavior could be drawn. The Pentium 4's instruction tagging mechanisms operate as follows. As the Pentium 4 decodes each instruction, it breaks it into a sequence of one or more simple operations called micro-operations or uops. The Pentium 4 tagging mechanisms enable these uops to be tagged when they cause certain performance events. Once a uop is tagged, it retains this tag until it retires successfully or is canceled. As the uops pass through the retirement logic, tagged uops that retire can be counted, providing a nonspeculative event count.

The accuracy of event-based sampling (EBS) support is also significantly improved in the Pentium 4. EBS support in previous processors typically monitored a particular performance counter and signaled a performance monitor interrupt (PMI) when the counter overflowed. The PMI ISR would then save the value of the program counter for the interrupted instruction. However, it is usually the case that the interrupted instruction is not the instruction that caused the performance counter to overflow. The processor's pipeline and the delay between counter overflow and the signaling of an interrupt combine to create an arbitrary distance (in retired instructions) between the instruction causing the performance event and the instruction whose address is saved by the PMI ISR. As such, EBS profiles on prior processors were often too inaccurate to be useful. In contrast, the Pentium 4 processor uses its uop tagging mechanisms along with microcode-level traps rather than macroinstruction-level interrupts to provide support for precise event-based sampling (PEBS). By recognizing tagged uops as they are about to retire and using a microcode-level trap to collect EBS data, the imprecision of prior EBS implementations is avoided because no instructions are allowed to retire between the retirement of the eventing instruction and the capturing of the sample data. The Pentium 4 EBS support also has a special buffering mechanism that automatically collects many samples before a macro-level interrupt service routine must be invoked to save the samples. This buffering technique significantly lowers the overhead associated with EBS, resulting in fewer perturbations of the application being monitored.

Because the Pentium 4 was also the first microprocessor to support simultaneous multithreading [11,10], its EMON capabilities support several thread qualification options. *Simultaneous multithreading (SMT)* is a technique that allows more than one task (thread) to share the processor's resources concurrently [16]. Thus, a single SMT processor appears to the operating system as multiple processors, and the operating system will execute ready-to-run threads for each processor it sees. The SMT approach provides a throughput advantage over a nonthreaded processor by allowing the processor's resources to continue to be used when one execution thread becomes stalled (e.g., due to a cache miss). The Pentium 4 processor provides SMT support by allowing its resources to be shared by two threads in dual-thread mode. The Pentium 4 can also operate in single-thread mode where the whole processor is allocated to only one thread. The Pentium 4 EMON hardware supports these SMT features by allowing event detection to be qualified by thread ID and thread mode. Each Pentium 4 event detector allows one to specify that event detection should only occur for thread 0, or thread 1, or for either thread. Each Pentium 4 event counter allows one to specify that event counting should occur when operating in single-thread mode, dual-thread mode, any-thread mode, or no-thread mode. In no-thread mode, neither thread is active, but the processor may still respond to bus transactions initiated by other active processors in the system.

A representative subset of the available Pentium 4 EMON events is presented in Figure 11.1. These events illustrate the type and diversity of the Pentium 4 events. The events in Figure 11.1 are divided into two categories: normal events and tag events. The normal events are selected by configuring individual event detectors and counters. The tag events are selected by configuring multiple event detectors (e.g., one that tags an instruction associated with a key event and another that detects tagged instructions as they complete execution). These and other events enable the performance analyst to determine the characteristics of an application (e.g., the types and counts of instructions used by the application) as well as how the Pentium 4 performs when executing the application (e.g., cache and TLB hits and misses, branch mispredictions, front-side bus traffic, and pipeline flushes).

11.3.2 The Pentium 4 EMON interface

The increased feature set and accuracy of the Pentium 4 EMON capabilities comes at the price of increased complexity. The interface between system software and Pentium 4 EMON hardware is significantly more complex than on prior processors. Even the simplest configuration to detect and count one event requires the setup of two EMON machine-specific registers (MSRs), an event detector, and an event counter. An EMON configuration for more than one event requires the careful allocation of event detectors and event counters, because only a subset of the 48 event detectors and 18 counters can be used to count a particular event. This section provides a brief overview of the Pentium 4 EMON interface, describing the organization of the event detectors and counters along with their configuration, the uop-tagging

```
Normal Event Name                  Brief Description of Events Detected
-----------------                  -----------------------------------
bpu_fetch_request                  trace cache misses
bsq_active_entries                 bus queue entries by type each cycle
branch_retired                     branches taken/not-taken and predicted/mispredicted
bsq_cache_reference                cache hits (exclusive, modified, shared) and misses
execution_event                    retiring uops tagged by the execution units
front_end_event                    retiring uops tagged by the front end
fsb_data_activity                  front-side bus activity (bus busy, bus ready)
global_power_events                cycles the processor is not stopped
instr_retired                      speculative, non-speculative, and tagged instructions retired
ioq_active_entries                 I/O queue entries by type each cycle
itlb_reference                     ITLB hits and misses
load_port_replay                   split loads
machine_clear                      events that clear the processor's pipeline
memory_cancel                      cancelled requests in the data cache
memory_complete                    completion of uncacheable and split loads and stores
mmx_128bit_uop                     128-bit MMX executions
mmx_64bit_uop                      64-bit MMX executions
mob_load_replay                    replayed loads
packed_dp_uop                      packed, double-precision executions
packed_sp_uop                      packed, single-precision executions
page_walk_type                     page walks for DTLB and ITLB misses
replay_event                       replayed uops tagged via the replay tagging mechanism
retired_branch_type                retiring branches by type: call, conditional, indirect, return
retired_mispred_branch_type        mispredicted retiring branches by type
scalar_dp_uop                      scalar, double-precision executions
scalar_sp_uop                      scalar, single-precision executions
sse_input_assist                   assists to handle input operand problems for SSE instructions
store_port_replay                  split stores
tc_deliver_mode                    cycles spent in build or deliver mode for the trace cache
tc_ms_xfer                         changes in uop delivery from the trace cache to the MS ROM
uop_queue_writes                   uop queue writes from the MS ROM, TC build, and TC deliver
uops_retired                       speculative and non-speculative uops retired
x87_assist                         floating point assists
x87_fp_uop                         floating point executions

Tag Event Name                     Brief Description of Uops Detected at Retirement
--------------                     ------------------------------------------------
dtlb_miss_retired                  loads that missed the DTLB
ld_miss_1L_retired                 loads that missed the 1st level cache
ld_miss_2L_retired                 loads that missed the 2nd level cache
loads_retired                      all loads
split_ld_retired                   split loads
split_st_retired                   split stores
stores_retired                     retiring stores
unaligned_ld_retired               loads with unaligned addresses
x87_fp_retired                     floating point uops
```

Figure 11.1 Example Pentium 4 EMON events.

mechanisms for obtaining nonspeculative event counts, and the PEBS support and sample buffering mechanism.

11.3.2.1   *Event detectors and counters*

The basic elements of the Pentium 4 EMON hardware are the event detectors and event counters. These are configured via two types of MSRs, the event select control register (ESCR) and the counter configuration control registers (CCCR). An ESCR is used to select the event to be detected, and the CCCR is used to configure the counter for the detected events. Each CCCR is also paired with a 40-bit counter that holds the count for the detected events. As mentioned previously, an ESCR can only detect a subset of the available EMON events, and the detected events from an ESCR can only be counted

BPU	ISTEER	IXLAT	ITLB	PMH	MOB	FSB	BSU
ESCR0	ESCR0	ESCR0	ESCR0	ESCR0	ESCR0	ESCR0	ESCR0
ESCR1	ESCR1	ESCR1	ESCR1	ESCR1	ESCR1	ESCR1	ESCR1
0	1	2	3	4	5	6	7

CCCR/Counter 0
CCCR/Counter 1
CCCR/Counter 2
CCCR/Counter 3
BPU

MS	TC	TBPU
ESCR0	ESCR0	ESCR0
ESCR1	ESCR1	ESCR1
0	1	2

CCCR/Counter 0
CCCR/Counter 1
CCCR/Counter 2
CCCR/Counter 3
MS

FLAME	FIRM	SAAT	U2L	DAC
ESCR0	ESCR0	ESCR0	ESCR0	ESCR0
ESCR1	ESCR1	ESCR1	ESCR1	ESCR1
0	1	2	3	4

CCCR/Counter 0
CCCR/Counter 1
CCCR/Counter 2
CCCR/Counter 3
FLAME

IQ	ALF	RAT	CRU		
ESCR0	ESCR0	ESCR0	ESCR0	ESCR2	ESCR4
ESCR1	ESCR1	ESCR1	ESCR1	ESCR3	ESCR5
0	1	2	4	5	6

ESCR Select Values For CCCRs

CCCR/Counter 0
CCCR/Counter 1
CCCR/Counter 4
CCCR/Counter 2
CCCR/Counter 3
CCCR/Counter 5
CRU

Figure 11.2 Pentium 4 ESCR and CCCR/counter interconnections.

on a subset of the available CCCR/counter pairs. Figure 11.2 shows the available interconnections between the ESCRs and CCCR/counter pairs. The names of the Pentium 4 functional units that contain the ESCRs and CCCR/counter pairs are noted in the top left of each box in Figure 11.2 (e.g., the BPU and TC). However, Intel has not published the full meaning and function of each unit (although some are easy to guess, such as BPU for the branch prediction unit and TC for the trace cache unit). Each gray box on the left side of Figure 11.2 represents a unit on the Pentium 4 processor that contains event detectors. To support the simultaneous counting of the same event from two different threads while running dual-thread mode, each of these units has two similarly equipped ESCRs (except for the CRU, which as three pairs of ESCRs). The units containing ESCRs are grouped into four sets shown horizontally in Figure 11.2. Each of these ESCR sets shares one group of CCCR/counter pairs, shown on the right side of Figure 11.2. The ESCR event select values indicated in the lower left of each unit box are used when configuring a CCCR to indicate which ESCR's events should be counted.

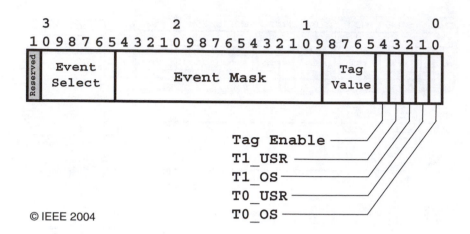

Figure 11.3 Pentium 4 event select control register (ESCR).

Each ESCR supports the selection of a particular event and allows the detection of that event to be qualified by an event-specific mask, thread ID, and privilege level. The layout of the ESCRs is shown in Figure 11.3. The ESCR event mask allows subsets of the event selected to be counted. For example, the instr_retired event may be further qualified by the selection of speculative instructions, nonspeculative instructions, tagged instructions, or nontagged instructions. The ESCR supports qualification of event detection by thread ID and privilege level using the T1_USR, T1_OS, T0_USR, and T0_OS bits. For example, to count only user-level events for thread one, only the T1_USR bit should be set. The tag enable and tag value fields of the ESCR are used by the execution tagging mechanism discussed in section 11.3.2.2.

The CCCR (see Figure 11.4) allows event counting to be qualified by thread mode and threshold and also supports several features related to counter overflow. Because the Pentium 4 supports SMT, the processor may execute in one of several SMT modes: single-thread mode, dual-thread mode, and no-thread mode. The CCCR supports qualification of event counting for each or any of these modes (single-thread, dual-thread, no-thread, or any-thread mode). The CCCR also allows event counting to be qualified by a threshold. The threshold support allows counting to occur when the number of events reported in one cycle from an event detector is greater-than or less-than-or-equal to the specified threshold. The compare, complement, edge, and threshold bit fields in the CCCR are used to configure the threshold support. The CCCR also contains an overflow flag bit (OVF) that is set when the counter overflows. The CCCR can be configured to request an interrupt on overflow (via the OVF_PMI_T0 and OVF_PMI_T1 bits) and also be forced to overflow on every event occurrence (via the Force_OVF bit). Finally, upon counter overflow, the CCCR can be configured to enable another CCCR to begin counting (via the Cascade bit).

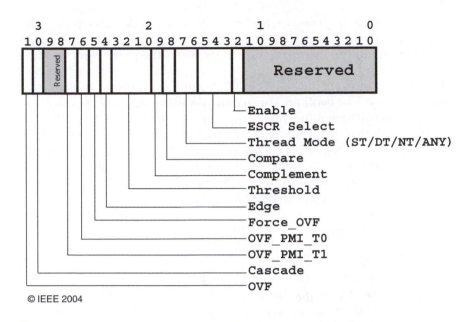

© IEEE 2004

Figure 11.4 Pentium 4 counter configuration control register (CCCR).

11.3.2.2 Tagging mechanisms for obtaining nonspeculative event counts

The Pentium 4 tagging mechanisms allow nonspeculative event counts to be obtained by enabling uops that encounter certain events to be tagged and counting tagged uops as they retire. The Pentium 4 implements three different tagging mechanisms: front-end, execution, and replay. The front-end tagging mechanism is able to tag uops responsible for events that occur early in the pipeline that are related to instruction fetching, instruction types, and uop delivery from the trace cache. The execution tagging mechanism is able to tag certain arithmetic uops as they write their results back to the register file. This mechanism uses a user-specified, 4-bit field to tag the uops (via the tag enable and the tag value fields in the ESCR). These uops can be tested as they retire and counted if any of these tag bits are set. The replay tagging mechanism is able to tag uops that are reissued (replayed) due to various conditions such as cache misses, branch mispredictions, dependence violations, and resource conflicts. The replay tagging mechanism relies upon two other MSRs that must be configured to select the desired replay cause (e.g., a cache miss) as well as the desired uop type (e.g., a load).

11.3.2.3 Event-based sampling support

The Pentium 4 support for PEBS provides two main advantages over prior processors: accuracy and low sampling overhead. Prior processors that relied solely upon an interrupt service routine to collect sample data upon performance counter overflow were inaccurate because often the data collected did not correspond to the instruction that caused the performance counter to overflow. Furthermore, an interrupt was needed to collect each sample, and frequent interrupts can significantly perturb the behavior of the system being monitored. The Pentium 4 overcomes these problems by using a microcode assist to collect sample data for tagged instructions as they retire and by buffering the samples collected. The microcode assist always collects the sample data from a tagged (an event-causing) instruction, so the information is precise. Each sample collected is placed directly in a memory buffer by the microcode assist, and only when the buffer is approaching capacity is an interrupt service routine invoked to harvest the samples.

The setup of the Pentium 4 PEBS support is complex. In addition to configuring the tagging mechanism for the desired EMON event, the Pentium 4 debug store (DS) area must be configured and a PEBS interrupt service routine must be installed. The DS configuration specifies the base, threshold, and maximum memory addresses for the PEBS buffer along with the buffer index and the reset value for the event counter. The PEBS interrupt service routine harvests the collected samples and re-enables the PEBS interrupt.

11.4   The brink and abyss tools

The advanced EMON capabilities of the Pentium 4 along with its aggressive processor design combine to make the task of understanding and using these EMON features significantly more difficult than with previous processors. This section describes the *brink* and *abyss* tools [13] that were designed to address these difficulties and to provide a high-level, easy-to-use interface for Pentium 4 EMON features on Linux systems. First, the structure and operation of the brink and abyss tools are described. Next, the brink user interface program is discussed along with its two key input files: the EMON configuration file and the experiment file. The abyss front-end program and device driver are then described followed by a brief example of the usage of the brink and abyss tools. This section closes with a discussion of the advantages of the brink and abyss approach.

11.4.1   The structure and operation of brink and abyss

The high-level structure and operation of the brink and abyss tools are illustrated in Figure 11.5. Initially, the brink user interface program reads the XML descriptions of the Pentium 4 EMON capabilities and the desired EMON setup and then creates a detailed EMON configuration for the desired experiments (i.e., the values for the various ESCRs, CCCRs, counters, and other EMON MSRs). Brink passes this detailed EMON configuration to the abyss front end, which passes it along to the abyss device driver. The abyss device driver uses the detailed EMON configuration to initialize the various EMON MSRs and, if necessary, set up an interrupt service routine that will

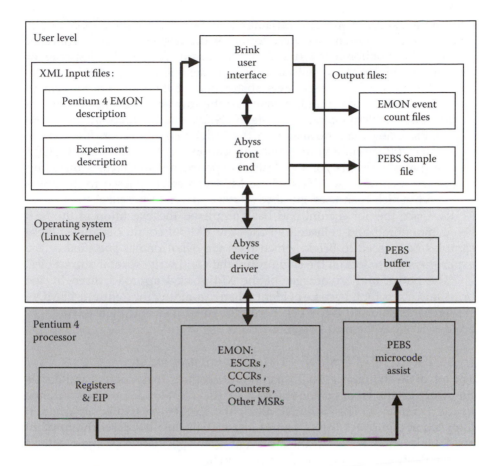

Figure 11.5 The structure of brink and abyss.

harvest samples from the PEBS buffer and pass them up to the abyss front end. Once the abyss front end has passed the detailed EMON configuration to the abyss device driver, it executes the application to be monitored, and periodically samples the EMON counters. The abyss front end buffers these EMON counter samples until the application completes and then they are passed to brink. Brink then processes the EMON counter sample data received from the abyss front end and writes several summary output files to disk. If PEBS is being used, the abyss front end also buffers these samples until the application being monitored completes and then writes a PEBS sample output file to disk.

11.4.2 The brink user interface program

The brink user interface program is a perl script that performs the majority of the work to make the Pentium 4 EMON capabilities easily accessible to the user. Brink expects two input files in XML format: a description of the

Pentium 4 EMON events and capabilities (referred to as the *EMON config-uration* file) and a high-level description of the desired Pentium 4 EMON events to be monitored (referred to as the *experiment* file). The experiment file relies upon the base events and symbols defined in the EMON configu-ration file to provide a high-level abstraction of the Pentium 4's EMON hardware, hiding from the user many of the low-level details, such as bit field positions, the mapping of events to ESCRs, and the mapping of ESCRs to CCCRs. Using the information in the EMON configuration file and the experiment file, brink determines an allocation for the ESCRs and CCCR/ counter pairs necessary for the desired experiments and creates the setup values for each of these MSRs. These MSR values are passed to the abyss front end, which then uses the abyss device driver to initialize the EMON MSRs. Once the abyss front end has completed the execution of the task being monitored and collected all the raw EMON counter samples, these samples are passed to brink, which uses the event names specified in the experiment file to format the event count data and write several output files.

The power and advantages of the XML-based approach used by the brink and abyss tools become evident when examining the Pentium 4 EMON configuration file and the experiment file. These files and their advantages are described in the next two subsections.

11.4.2.1   The Pentium 4 EMON configuration file

The job of the Pentium 4 EMON configuration file is to document and define the structure and layout of the ESCRs, CCCRs, counters, and events available in the Pentium 4. This includes all of the ESCR and CCCR information depicted in Figures 11.1, 11.2, and 11.3 as well as detailed descriptions of all available events, including the *tag* events (nonspeculative events obtained via the Pentium 4 uop tagging mechanisms).

An XML representation for the EMON configuration file was chosen for the following reasons. The current approach for documenting a processor's EMON capabilities is to describe them in a reference manual [5]. This approach, although good and sufficient, requires programmers to translate the verbal and pictorial description of these capabilities from the reference manual into data structures for their programs. Furthermore, when new versions of the processor are released with additional EMON capabilities and/or events, these additions must be noted by the programmer, the cor-responding data structures must be updated, and new releases of the soft-ware must be made. This reference-manual approach would be significantly improved if the description of the EMON capabilities were not just human-readable but also directly usable by the software tools that employ the EMON capabilities.

Among the key design goals for XML documents were the following: XML documents should be easy to create, XML documents should be human-readable, and it should be easy to write programs that process XML documents [17]. As such, an XML description for the EMON capabilities is an excellent choice. The human-readable quality of an XML description of

```
<!-- CCCR Configuration -->
<cccr_config size="64">
    <reserved1    bits="0-11"   default="000000000000"/>
    <enable       bits="12"     default="1"/>
    <escr_select  bits="13-15"  default="000"/>
    <reserved2    bits="16-17"  default="11"/>
    <compare      bits="18"     default="0"/>
    <complement   bits="19"     default="0"/>
    <threshold    bits="20-23"  default="0000"/>
    <edge         bits="24"     default="0"/>
    <force_ovf    bits="25"     default="0"/>
    <ovf_pmi      bits="26"     default="0"/>
    <reserved3    bits="27-29"  default="000"/>
    <cascade      bits="30"     default="0"/>
    <ovf          bits="31"     default="0"/>
    <reserved4    bits="32-63"  default="00000000000000000000000000000000"/>
</cccr_config>

<!-- ESCR Configuration -->
<escr_config size="64">
    <reserved1    bits="0-1"    default="00"/>
    <usr          bits="2"      default="0"/>
    <os           bits="3"      default="0"/>
    <tag_enable   bits="4"      default="0"/>
    <tag0         bits="5"      default="0"/>
    <tag1         bits="6"      default="0"/>
    <tag2         bits="7"      default="0"/>
    <tag3         bits="8"      default="0"/>
    <event_mask   bits="9-24"   default="0000000000000000"/>
    <event_select bits="25-30"  default="000000"/>
    <reserved2    bits="31"     default="0"/>
    <reserved3    bits="32-63"  default="00000000000000000000000000000000"/>
</escr_config>

<!-- Counter Configuration -->
<counter_config size="64">
    <counter      bits="0-39"   default="0000000000000000000000000000000000000000"/>
    <reserved     bits="40-63"  default="000000000000000000000000"/>
</counter_config>
```

Figure 11.6 ESCR, CCCR, and counter definitions.

EMON capabilities augments the reference manual description, improving the understanding of the programmers using the EMON capabilities, while also providing a directly usable description of the essential data structures needed to write programs that use the EMON capabilities. The human-readable quality of an XML description also makes the updating of the EMON capabilities easier as new events and/or features are implemented in newer versions of the processor. Furthermore, because of the widespread success and use of XML, various parsers are readily available for XML documents.

An excerpt from the EMON configuration file that contains the XML description of the layout of the Pentium 4 ESCRs, CCCRs, and counters is shown in Figure 11.6. The name and range of each bit field in these registers is defined along with a default value. The names for the bit fields can be used in the experiment file to define specific values for the fields; otherwise the default value will be used.

An excerpt from the EMON configuration file, shown in Figure 11.7, contains the XML description of the mapping and interconnections between the ESCRs, CCCRs, and counters depicted in Figure 11.2. Note that, although only a subset of the definitions are shown in Figure 11.7, the omitted definitions are similar in structure to the ones shown. First, the set of CCCRs is

```
<!-- CCCR Names and Addresses -->
<cccr>
    <bpu_cccr0      address="0x360" counter="bpu_counter0"/>
    <bpu_cccr1      address="0x361" counter="bpu_counter1"/>
    <bpu_cccr2      address="0x362" counter="bpu_counter2"/>
    <bpu_cccr3      address="0x363" counter="bpu_counter3"/>

    <flame_cccr0    address="0x368" counter="flame_counter0"/>
    <flame_cccr1    address="0x369" counter="flame_counter1"/>
    <flame_cccr2    address="0x36A" counter="flame_counter2"/>
    <flame_cccr3    address="0x36B" counter="flame_counter3"/>

    <!- the remaining definitions omitted here for brevity -->
</cccr>

<!-- Counter Names and Addresses -->
<counter>
    <bpu_counter0   address="0x300" rdpmc_index="0"/>
    <bpu_counter1   address="0x301" rdpmc_index="1"/>
    <bpu_counter2   address="0x302" rdpmc_index="2"/>
    <bpu_counter3   address="0x303" rdpmc_index="3"/>

    <flame_counter0 address="0x308" rdpmc_index="8"/>
    <flame_counter1 address="0x309" rdpmc_index="9"/>
    <flame_counter2 address="0x30A" rdpmc_index="10"/>
    <flame_counter3 address="0x30B" rdpmc_index="11"/>

    <!- the remaining definitions omitted here for brevity -->
</counter>

<!-- ESCR Names and Addresses -->
<escr>
    <fsb_escr0 address="0x3a2" select="110">
        <cccr>bpu_cccr0</cccr>
        <cccr>bpu_cccr1</cccr>
    </fsb_escr0>

    <fsb_escr1 address="0x3a3" select="110">
        <cccr>bpu_cccr2</cccr>
        <cccr>bpu_cccr3</cccr>
    </fsb_escr1>

    <dac_escr0 address="0x3a8" select="101">
        <cccr>flame_cccr0</cccr>
        <cccr>flame_cccr1</cccr>
    </dac_escr0>

    <dac_escr1 address="0x3a9" select="101">
        <cccr>flame_cccr2</cccr>
        <cccr>flame_cccr3</cccr>
    </dac_escr1>

    <!- the remaining definitions omitted here for brevity -->
</escr>
```

Figure 11.7 Mapping definitions for ESCRs, CCCRs, and counters.

defined and, for each CCCR, the MSR address and associated counter are defined. Second, the set of counters is defined and, for each counter, the MSR address and RDPMC index are defined (RDPMC is an instruction that reads the value of a counter). Third, the set of ESCRs is defined and, for each ESCR, the MSR address and the select value used when configuring a CCCR to count the ESCR's events are defined. Also defined for each ESCR is a list that indicates all of the CCCRs that can be configured to count the ESCR's events. For example, the event detectors configured via the fsb_escr0 ESCR can be counted only by the counters controlled by the bpu_cccr0 and

```
<normal_events>

    <retired_branch_type>
        <event_select>0x04</event_select>
        <event_mask>
            <conditional bits="10"/>
            <call        bits="11"/>
            <return      bits="12"/>
            <indirect    bits="13"/>
        </event_mask>
        <escr>tbpu_escr0</escr>
        <escr>tbpu_escr1</escr>
    </retired_branch_type>

    <itlb_reference>
        <event_select>0x18</event_select>
        <event_mask>
            <hit    bits="9"/>
            <miss   bits="10"/>
            <hit_uc bits="11"/>
        </event_mask>
        <escr>itlb_escr0</escr>
        <escr>itlb_escr1</escr>
    </itlb_reference>

<!-- the remaining definitions omitted here for brevity -->

</normal_events>
```

Figure 11.8 Examples of normal event definitions.

bpu_cccr1 CCCRs, whereas the event detectors configured via the fsb_escr1 ESCR can be counted only by the counters controlled by the bpu_cccr2 and bpu_cccr2 CCCRs.

Once the layout and interconnections between the ESCRs, CCCRs, and counters have been defined, it is fairly straightforward to create XML descriptions of the Pentium 4 EMON events. Two examples of normal (i.e., not tag) event definitions are shown in Figure 11.8. The first example shown in Figure 11.8 is for the retired branch type event, retired_branch_type. First, the ESCR event select field value used to select this event is given (0x04). Next, the four event mask bits that allow types of retiring branches to be counted are defined (conditional, call, return, and indirect). As the last part of the event definition, a list of the ESCRs that can be used to select this event is given (ESCRs tbpu_escr0 and tbpu_escr1). Note that the structure of the second example shown in Figure 11.8 for the itlb_reference event is very similar in structure to the definition for the retired_branch_type event. In this manner, all of the normal Pentium 4 events are defined in the EMON configuration file.

Once the normal events have been defined, the tag event definitions, which rely upon combinations of normal events and the Pentium 4 tagging mechanisms, are defined. Two examples of tag event definitions are shown in Figure 11.9. The tag event definitions are composed of two parts: the tag setup and the count setup. The tag setup defines the event that will be used

```
<tag_events>

    <loads_retired>
        <type>front_end</type>
        <tag_setup>
            <uop_type>
                <set><tagloads/></set>
            </uop_type>
        </tag_setup>
        <count_setup>
            <front_end_event>
                <set><nbogus/></set>
            </front_end_event>
        </count_setup>
    </loads_retired>

    <unaligned_ld_retired>
        <type>replay</type>
        <tag_setup>
            <mob_load_replay>
                <set>
                    <partial_data/>
                    <unalgn_addr/>
                </set>
            </mob_load_replay>
            <pebs_enable>
                <set>
                    <mob/>
                    <tag/>
                </set>
            </pebs_enable>
            <pebs_matrix_vert>
                <set>
                    <ld/>
                </set>
            </pebs_matrix_vert>
        </tag_setup>
        <count_setup>
            <replay_event>
                <set><nbogus/></set>
            </replay_event>
        </count_setup>
    </unaligned_ld_retired>

    <!-- the remaining definitions omitted here for brevity -->

</tag_events>
```

Figure 11.9 Examples of tag event definitions.

to tag a uop as it flows through the Pentium 4 pipeline. The count setup defines the event that will be used to count the tagged uops as they exit the pipeline.

Figure 11.9 shows two examples of tag event definitions. The first example in the upper half of Figure 11.9 is for the load_retired tag event that uses the front-end tagging mechanism. The tag setup for this event uses the uop_type event to tag the load uops as they pass through the front end of the pipeline. This is specified by setting the tagloads bit in the event mask for the uop_type event. The count setup for this event uses the front_end_event event with the nbogus bit set (a nonbogus uop is a uop that is along the correct program execution path). The second example in lower half of Figure 11.9 shows the definition for one of the more complex tag events, unaligned_ld_retired. This tag setup for this event requires three different things to be configured in order to tag the load uops that have unaligned accesses. First, the mob_load_replay must be enabled with the partial_data and unalgn_addr bits set in the event mask. Next, two other special MSRs must be configured—pebs_enable and pebs_matrix_vert—to indicate that the Pentium 4 MOB (memory order buffer) should tag load uops that are replayed due to unaligned addresses and/or partial data in the MOB. The count setup for this event uses the replay_event event with the nbogus bit set. Note that the XML-based approach and the brink user interface program are flexible enough to accommodate an arbitrary number of actions for the tag setup.

11.4.2.2 The Experiment File

The job of the XML-based experiment file is to allow the user to specify the programs to be monitored along with the desired EMON event configuration to the brink user interface program. The event names and bit fields defined in the EMON configuration file are used in the experiment file to select the desired EMON events to monitor and define their configuration. Although many EMON MSRs must be configured to collect EMON data, no specific EMON MSRs (e.g., which ESCRs to use) must be specified in the experiment file. The brink and abyss tools allocate these EMON MSRs without user intervention, greatly simplifying the configuration task for the user. The brink and abyss tools will execute a number of jobs equal to the product of the number of applications and the number of experiments defined to collect the desired EMON data. As can be seen in example experiment file shown in Figure 11.10, an experiment file is composed of two sections: the programs section and the experiments section.

The *programs* section of the experiment file defines the list of the programs to be monitored along with the invocation command for the programs. The experiment file shown in Figure 11.10 lists two simple benchmark programs: inst_10B, which executes 10 billion instructions, and unalign_ld, which executes many unaligned load instructions.

The *experiments* section of the experiment file shown in Figure 11.10 describes the setup for two separate groups of EMON event configurations, referred to as experiments. Separate experiment descriptions are often necessary because many EMON events share the same event detectors and counters, and therefore require separate application runs to collect all the desired event count data. The first part of the experiments section describes the default settings for all the experiment configurations. The default section shown in Figure 11.10 enables event detection for both USR and OS modes. The next two parts of the experiments section define the uop and PEBS experiments.

The uop experiment shown in Figure 11.10 describes the EMON configuration necessary to count all nonbogus uops that retire as well as the number of cycles in which three nonbogus uops retire. In the uop experiment, two new event names are defined: uops_retired_all and three_uop_retire_cycles. These event names will be used in the various output files when referring to these event counts. The specification of base="uops_retired" in these event definitions selects the uops_retired event defined in the EMON configuration

```
<exp_config>
    <processor type="pentium4"/>

    <programs>
        <inst_10B   command="./progs/inst_10B"/>
        <unalign_ld command="./progs/unalign_ld"/>
    </programs>

    <experiments>

        <default>
            <set>
                <usr/> <!-- detect USER mode events -->
                <os/>  <!-- detect OS mode events -->
            </set>
        </default>

        <uop_experiment>
            <uops_retired_all base="uops_retired">
                <set>
                    <nbogus/>      <!-- detect non-bogus (real) uops -->
                </set>
            </uops_retired_all>

            <three_uop_retire_cycles base="uops_retired">
                <set>
                    <nbogus/>     <!-- detect non-bogus (real) uops -->
                    <compare/>    <!-- compare with threshold value -->
                </set>
                <threshold val="0010"/> <!-- only occurrences > 2 will be counted -->
            </three_uop_retire_cycles>
        </uop_experiment>

        <pebs_experiment>
            <unaligned_ld_retired_usr base="unaligned_ld_retired">

                <clr><os/></clr>  <!-- detect only USR mode events -->

                <ebs type="precise" interval="100000" buffer="500" sample="E" max="20000"/>

            </unaligned_ld_retired_usr>
        </pebs_experiment>

    </experiments>
</exp_config>
```

Figure 11.10 An example experiment configuration file for brink.

file to serve as the base event for this event definition. The set directive for the nbogus bit in these event definitions indicates that the nbogus bit in the event mask should be set to enable the detection of all nonbogus retiring uops. The set directive for the compare bit in the three_uop_retire_cycles event indicates that the number of events detected each cycle should be compared to a threshold value and that the counter should be incremented only if the number of events detected exceeds the threshold. The val directive for the threshold field specifies a threshold value of two (binary 0010), so only cycles where three uops retire will cause the counter to be incremented.

The PEBS experiment shown in Figure 11.10 describes the EMON configuration necessary to configure PEBS for unaligned loads that retire. The definition of the unaligned_ld_retired_usr event specifies that the unaligned_ld_retired event defined in the EMON configuration file (and shown in Figure 11.9) serves as the base event for this definition. The clr directive for the OS bit for this event definition specifies that no events in OS mode (i.e., when the operating system is executing) should be detected. The ebs directive for this event specifies the necessary parameters to configure the abyss device driver to setup PEBS and collect the resulting samples. This ebs directive indicates that precise samples be collected (as opposed to collecting imprecise samples by generating interrupts on counter overflows), that the interval between each sample should be 100,000, that 500 samples should be buffered before copying the samples from kernel space to user space, and that a maximum of 20,000 samples be collected.

As can be seen from the examples discussed earlier, the choice to use XML for the experiment file provides a number of advantages. First, the user can define the event names to be used for each specific event configuration, avoiding the necessity to predefine all possible event names or force the use of cumbersome event-naming conventions. Second, the user does not have to allocate the ESCRs, CCCRs, and counters for each event and does not have to know or use the various MSR addresses, event select encodings, and ESCR selection values. Third, the configuration of the event detection and counting options is done symbolically, using the symbols defined in the EMON configuration file, eliminating the need to keep track of the width and location of various bit fields. Fourth, many application runs to collect various sets of event data can all be specified in one experiment file, automating the collection of large amounts of data. Fifth, all of the advantages cited earlier for choosing XML for the EMON configuration file also apply to the experiment file: The experiment file is human-readable, easy to create, and easy to process.

11.4.3   The abyss front-end program and device driver

The low-level job of initializing the EMON hardware is handled by the abyss front-end program and device driver. For each job to be run (a job is the combination of one application and one EMON experiment configuration), the brink user interface creates a detailed description of the EMON hardware configuration and the application to monitor and passes this to the abyss front end, which interacts directly with the abyss device driver to configure the EMON hardware and collect event count and PEBS samples. When executed, the abyss front end creates additional processes to execute the applications that will be monitored and, if EBS is enabled, to collect event-based samples. The execution of these processes is coordinated by the abyss front end with help from the abyss device driver. The abyss device driver is a Linux device driver module that provides user-level access to the Pentium 4 EMON MSRs via read(), write(), and ioctl() calls to the /dev/ abyss device file. In addition to providing access to the EMON MSRs, the abyss device driver also configures the Pentium 4's event-based sampling features by allocating the PEBS buffer in the kernel and creating the interrupt service routines for the imprecise and precise event-based sampling interrupts. Upon the completion of the job, the abyss front end passes the event count samples back to the brink user interface program, which then creates a collection of data files containing summaries of the job setup and event counts along with a set of detailed data files containing all the raw sample data. Unlike brink, which is a Perl script, the abyss front-end program and device driver are written in C to minimize the perturbation of the applications being monitored as event count and PEBS samples are collected.

11.4.4   A Brink and Abyss Example

When brink is run using the experiment file shown in Figure 11.10, a number of output files are created in an emon_data directory. These files include copies of the experiment and EMON configuration files along with a directory for each job executed. For example, four job directories will be created when the experiment file shown in Figure 11.10 is used (two programs to execute, each using two different EMON configurations). The brink user interface program also creates a summary file in the emon_data directory, which summarizes both the EMON MSR setup for each experiment and the event counts from each job executed.

Each job executed by brink produces a set of output files. First, a copy of the input file for the abyss front end is saved. Second, the raw, delta, and EBS sample data files are written. The raw data file contains the raw counter values obtained each time the abyss front end samples the performance counters. The delta file contains the differences between each of the performance counter samples contained in the raw data file. The EBS sample file contains the event-based samples collected by the abyss device driver when event-based sampling is enabled. These event-based samples are essentially the Pentium 4 register values, including the program counter, for each eventing instruction that was sampled. Finally, a summary file similar to the one discussed earlier and shown in Figure 11.11 is included in the job directory that details the EMON MSR setup and event counts just for the job.

An excerpt from the brink summary file resulting from the jobs run for the experiment file shown in Figure 11.10 is shown in Figure 11.11. The excerpt shown in Figure 11.11 only displays the information from the inst_10B program and the uop_experiment experiment. The summary file lists the names of the programs and experiments that were defined and, for each experiment, shows the EMON MSR configuration used to obtain the desired event counts. The uop_experiment experiment defined two events: uops_retired and three_uop_retire_cycles. The binary values for ESCR, CCCR, and performance counter for each of these events is shown. The summary file closes with a list of the total event counts for each event defined as well as the total cycles obtained from the time-stamp counter. For example, the total event counts displayed in Figure 11.10 show that the inst_10B program executed approximately 16 billion uops in 6.5 billion cycles and was able to retire three uops per cycle on 3.6 billion of the 6.5 billion cycles executed.

11.4.5   The Advantages of the brink and abyss approach

These XML descriptions of the Pentium 4's EMON features and the desired EMON configuration provide a number of advantages over the traditional, reference-manual approach for documenting performance monitoring features. First, the XML EMON feature description (i.e., the EMON configuration file)

```
BRINK SUMMARY FILE

User:        bsprunt
Host:        bsprunt-4.eg.bucknell.edu
Date:        Tue May 18 16:25:18 EDT 2004

Command:     /brink -exp example_exp.txt

Programs:
             inst_10B:    ./progs/inst_10B

Experiments:
             uop_experiment

Jobs:
             inst_10B.uop_experiment

MSR Setup For Each Experiment:

  uop_experiment:
    three_uop_retire_cycles:
      0x3b8 00000000000000000000000000000000000001000000000000000001000001100 cru_escr0
      0x36C 00000000000000000000000000000000000000000100111100100000000000000 iq_cccr0
      0x30C 00000000000000000000000000000000000000000000000000000000000000000 iq_counter0

    uops_retired_all:
      0x3b9 00000000000000000000000000000000000001000000000000000001000001100 cru_escr1
      0x36E 00000000000000000000000000000000000000000000011100100000000000000 iq_cccr2
      0x30E 00000000000000000000000000000000000000000000000000000000000000000 iq_counter2

LOG:

inst_10B.uop_experiment
  inst_10B.uop_experiment            time_stamp_counter           6,523,666,168
  inst_10B.uop_experiment            three_uop_retire_cycles      3,606,859,638
  inst_10B.uop_experiment            uops_retired_all            16,005,449,923
```

Figure 11.11 An excerpt from an example brink summary file.

provides a well-structured, human-readable document that augments the reference-manual documentation that can easily be updated or extended as new events and capabilities are implemented in successive versions of the Pentium 4 processor. Second, the XML EMON feature description can be directly used by application software to create the data structures necessary to manipulate the EMON hardware and allocate EMON resources. Third, the XML description of the desired EMON configuration (i.e., the experiment file) can be easily created by a user to specify desired performance monitoring configurations. Fourth, these performance-monitoring configurations are described symbolically, eliminating the need for the user to recall detailed EMON MSR addresses, layouts, and interrelationships. In summary, the XML-based approach used by brink and abyss to describe EMON features and configurations capitalizes upon three key benefits of XML documents (XML documents are human-readable, easy to create, and easy to process) to bring the power and flexibility of XML to the problem of improving the documentation and ease-of-use of the complex performance-monitoring hardware found in modern, high-performance microprocessors.

If future versions of performance-monitoring hardware were to be documented by the computer manufacturer in a fashion similar to the way the Pentium 4 features were documented using XML for the brink and abyss tools, then applications that exploit these features would be easier to develop and maintain. Also, the human-readable qualities of XML-based documentation would aid new users in developing a good understanding of the EMON features themselves.

11.5   Performance-monitoring issues and opportunities for future processors, systems, and software

Mainstream computer systems will soon be moving toward a greater use of symmetric multiprocessing (SMP) systems. Several major microprocessor manufacturers (IBM, Intel, HP, and Sun) have begun selling or will soon sell microprocessors that employ chip multiprocessing (CMP) in which more than one processor core is integrated into the processor package. Additionally, these manufacturers are also selling or will soon sell processors that support simultaneous multithreading (SMT), a technique that allows one physical processor core to be concurrently shared by more than one task. Consequently, typical computer systems and applications may soon be concurrently executing multiple-process applications on systems with multiple, multithreaded processors. This shift from uniprocessor, single-threaded systems toward multiprocessor, multithreaded systems has a number of implications for performance-monitoring hardware and its associated software tools.

A number of software tools are available to configure EMON hardware and collect performance data [2,9,3,1], but many of these tools assume that the applications being monitored consist of single tasks running on single-processor systems. This assumption allows the tools to configure the EMON hardware and collect performance counter data for the whole system rather than for each task. As such, a performance analyst typically will setup the system and application to be monitored such that only the desired application is executing (along with the operating system). With this setup, the system's performance is essentially equivalent to the application's performance. Following this approach simplifies the EMON software tools by eliminating the need to setup and track EMON configurations on a per-task or per-thread basis and eliminates the need to manage multiple sets of EMON hardware in one system (one set for each physical processor).

However, with widespread adoption of CMP and/or SMT processors running multithreaded applications on SMP operating systems, it will no longer be practical to assume that the performance of an application can be measured on a single-processor, single-threaded system as described. As such, future EMON software tools and systems will need to become more elaborate in order to track performance of individual tasks or collections of tasks in a per-task fashion on systems that support multiple, multithreaded processors. Creating these tools will require even closer interaction and coordination with the underlying operating system in order to associate performance event counts with the appropriate tasks or threads. As such, these tools are likely to be developed first on open-source operating systems, such as Linux.

As EMON hardware capabilities and software tools begin to accommodate the move to systems composed of CMP/SMT processors, additional opportunities for improving system performance lie in the area of operating system task scheduling. Because CMP/SMT processors by definition concurrently share processor and system resources among tasks, it may often be the case that some sets of tasks will share the resources more productively than others. For example, a multithreaded processor may productively execute two tasks concurrently that require different processor resources (e.g., one task that needs integer execution units and one task that needs floating-point execution units) but may not execute as productively two tasks that require the same processor resources (e.g., two tasks that both need the floating-point execution units). If the number of tasks that are ready to execute is greater than the number of logical or physical processors available in the system (as is often the case for servers), then the operating system may improve overall system performance (throughput) by selecting sets of tasks to run concurrently that productively share the system's resources. To do this type of scheduling, the operating system needs to obtain information regarding the performance of various task set combinations. The performance data needed by the operating system to make these high-level scheduling decisions can be obtained from EMON hardware. This type of scheduling has been shown to provide significant performance improvements for multithreaded processors and is referred to as symbiotic task scheduling [12]. As processors and systems move to greater levels of concurrency, and performance-monitoring hardware and tools accommodate these changes, the development of operating system schedulers that use dynamic processor performance data to guide task-scheduling decisions will become prevalent.

11.6   Summary

This chapter introduced the basic concepts behind the on-chip hardware used for performance event monitoring (referred to as *EMON* for *event monitoring*) in modern, high-performance microprocessors. The use of EMON detectors and counters to obtain program characterization and microprocessor performance data (such as instruction counts, cache misses, and branch mispredictions) was described. Additionally, the use of event-based sampling techniques to locate the code sequences responsible for key performance problems was discussed. The specific features of the Pentium 4's EMON hardware along with its improvements over prior processors were then introduced. Next, the brink and abyss tools that provide a high-level interface to the Pentium 4's EMON features were described. These tools employ the unique approach of using XML documents to describe the Pentium 4's EMON capabilities. This XML-based approach represents a significant improvement over the current, reference-manual approach for documenting EMON capabilities because, in addition to being human-readable (as a reference manual is), the XML description of the Pentium 4's EMON capabilities can also be directly used by EMON software tools to create the data structures necessary to manipulate the EMON configuration. In closing, this chapter discussed the issues and opportunities that arise for the users of EMON

hardware and developers of EMON software tools as mainstream microprocessors begin to employ simultaneous multithreading and/or chip multiprocessing techniques.

References

1. Berrendorf, Rudolf, Ziegler, Heinz, and Mohr, Bernd, *PCL, The Performance Counter Library*, online at: http://www.fz-juelich.de/zam/PCL/.
2. Browne, S., Dongarra, J., Garner, N., Ho, G., and Mucci, P., A portabe programming interface for performance evaluation on modern processors, *International Journal of High Performance Computing Applications* 14, 4, 189–204, 2000.
3. Heller, Don, *Rabbit, A Performance Counters Library for Intel/AMD Processors and Linux*, online at: http://www.scl.ameslab.gov/Projects/Rabbit/.
4. Hinton, Glenn, Sager, Dave, Upton, Mike, Boggs, Darrell, Carmean, Doug, Kyker, Alan, and Roussel, Patrice. The microarchitecture of the Pentium 4 processor. *Intel Technology Journal* 5, 01, February 2001, online at: http://www.intel.com/technology/itj/q12001/pdf/art 2.pdf.
5. Intel, *Intel Pentium 4 Processor Manuals*, online at: http://developer.intel.com/design/Pentium4/documentation.htm.
6. Intel, *Intel Pentium III Processor Manuals*, online at: http://developer.intel.com/design/PentiumIII/documentation.htm.
7. Intel, *Intel Pentium Pro Processor Manuals*, online at: http://developer.intel.com/design/archives/processors/pro/.
8. Intel, *Intel Pentium Processor Manuals*, online at: http://developer.intel.com/design/Pentium/manuals/.
9. Intel Corporation, *VTune Performance Analyzers*, online at: http://www.intel.com/software/products/vtune/index.htm.
10. Koufaty, David, and Marr, Deborah T., Hyperthreading technology in the Netburst microarchitecture, *IEEE Micro Magazine* 23, 2, 56–64, 2003.
11. Marr, D., Binns, F., Hill, D., Hinton, G., Koufaty, D., Miller, J., and Upton, M., Hyper-threading technology architecture and microarchitecture, *Intel Technology Journal* 6, 01, February 2002, online at: http://developer.intel.com/technology/itj/2002/volume06issue01/art01 hyper/vol6iss1 art01.pdf.
12. Snavely, Allan, and Tullsen, Dean M., Symbiotic jobscheduling for a simultaneous multithreading processor, in *Ninth International Conference on Architectural Support for Programming Languages and Operating Systems*, November 2000, online at: http://www-cse.ucsd.edu/users/tullsen/asplos00.pdf.
13. Sprunt, Brinkley. Brink and Abyss, *Pentium 4 Performance Counter Tools for Linux*, online at: http://www.eg.bucknell.edu/~bsprunt/emon/brink_abyss/brink_abyss.shtm.
14. Sprunt, Brinkley. Pentium 4 Performance Monitoring Features. *IEEE Micro Magazine*, 22(4):72–82, July-August 2002.
15. Sprunt, Brinkley, The basics of performance monitoring hardware, *IEEE Micro Magazine* 22, 4, 64–71, 2002.
16. Tullsen, D.M., Eggers, S.J., and Levy, H.M., Simultaneous multithreading: Maximizing on-chip parallelism, in *22nd Annual International Symposium on Computer Architecture*, June 1995, online at: http://www-cse.ucsd.edu/users/tullsen/ISCA95.ps.
17. W3C. *Extensible Markup Language (XML) 1.0 (Second Edition)*, online at: http://www.w3.org/TR/2000/REC-xml-20001006.

Chapter Twelve

Performance Monitoring on the POWER5™ Microprocessor

Alex Mericas

Contents

12.1   Introduction

Like most modern microprocessors, the IBM PowerPC Family of microprocessors incorporate built-in hardware for diagnostic and performance monitoring [1]. The POWER5 Performance Monitor Unit (PMU) is the result of
evolutionary improvements that started with the PowerPC 604™ Microprocessor. This chapter will explore the performance-monitoring capabilities of
the PowerPC, with emphasis on the POWER5 processor. Advanced features
of the POWER5 PMU allow performance problems to be quickly identified
and located with minimal overhead. Designed with flexibility in mind, the
PMU supports system-wide global monitoring and thread-level detailed
monitoring. In addition to the hardware capabilities, an overview of software
support for these facilities will also be discussed.

The POWER5 is the latest generation of the PowerPC™ family of processors from IBM. Like the POWER4, the POWER5 is a superscalar processor
that exploits speculative and out-of-order execution to minimize the performance impact of memory latency. Also like the POWER4, it has two physical
CPUs per chip. In additional to multiple processors per chip, the POWER5
uses simultaneous multithreading (SMT), making a single CPU appear as
two logical CPUs to the operating system [2]. Each CPU can run two independent threads, dynamically assigning resources as needed to optimize
throughput. Continuing an evolution that began with the PowerPC 604, the
POWER5 uses an integrated PMU to monitor, record, and report key performance indicators. Each thread has a dedicated PMU Unit that can be
independently configured to monitor the performance of the processor or
memory system.

12.1.1   Evolution of powerPC performance monitoring [3]

12.2   Basic features

The PowerPC performance monitor is register based; all control of the
counters and the counters themselves are *special-purpose registers (SPRs)*.
These SPRs are accessed through the *move to SPR (MTSPR)* and *move from
SPR (MFSPR)* instructions. The registers may be read in user mode but
require supervisor mode to be written to.

12.2.1   Counters

The *performance monitor counters (PMCs)* are 32-bit-wide SPRs. The width
of the counter determines how many event occurrences can be counted
before the counters overflow or wrap. The counters can be configured to
signal the operating system through a performance monitor exception
when the left-most (high order in PowerPC terminology) bit transitions

from 0 to 1, indicating that an overflow is imminent. Alternatively, one counter can be configured to overflow into another counter, chaining two or more counters together to increase the number of bits. For most events, 32 bits are more than sufficient even with higher-frequency processors. Because the counters cause an exception long before they actually wrap, the counters can continue to count, and the operating system is able to service the exception before the counters advance to the point of losing data. Of all the events counted, processor cycles are the most sensitive to counter overflow. The interrupt rate caused by the cycle counter is still much lower than other sources of interrupts, particularly the clock interrupt. There is no benefit of increasing the width of the counters.

As shown in Figure 12.1 the number of counters implemented varies by processor family (and in some cases within a family) from two to eight PMCs per processor. The PPC604 had two counters, the PPC604e added two more. The POWER3 and POWER4 implemented eight counters per CPU. The RS64 processor, which introduced hardware multithreading, supports eight PMCs but allows them to be shared between the two threads. The PMU is split into two sets of four counters in multithread mode with a common set of control registers; each set of counters monitor the same events for the two threads. Similarly, the Intel Pentium 4 allows any counter to count from either thread, or both combined. On POWER5 each thread appears to the operating system as a separate processor capable of running programs from different users independent of each other. The shared PMU approach was considered unworkable in this environment. Instead, POWER5 implements thread-specific PMUs with each PMU's counters independent of the other thread's. The ideal situation for POWER5 would have been to replicate the POWER4 PMU and double the number of counters to eight per thread, or 16 total. Total chip area was a concern, which led to an initial design with four counters per thread. Because most measurements include instructions and cycles a compromise between area and function was reached, adding two nonprogrammable counters to each PMU to count instructions and cycles.

The PMCs can be read directly without a state change (into supervisor state, for example). They can be read before and after code being measured for low overhead collection.

12.2.2 Control registers

The PowerPC PMU is controlled through SPRs known as *monitor mode control registers (MMCRs)*. The MMCRs control what events are monitored; when to start, stop, or pause (freeze) the counters; and other features of the PMU. The PPC604 had two 32-bit MMCRs, which was sufficient to control the two counters with bits to spare. As the PowerPC family evolved, these spare bits were quickly used up, and a third MMCR was defined. The bit fields within particular MMCRs were generally related by implementation date more than

PowerPC 604
> Two counters

PowerPC 604e™
> Four counters
> Load/Store Sampling

RS64™
> Eight counters
> Instruction Matching
> Hardware Multi-threading
> Continuous Sampling

POWER3™
> Eight counters
> Slot-based Sampling

POWER4™
> Eight counters
> Instruction Matching (multiple match conditions)
> Random Sampling
> Memory Source Feedback

PowerPC 970™
> Eight counters
> Completion Stall Accounting

POWER5™
> Twelve counters (6*2)
> Virtualization support
> Latency detection
> Completion Stall Accounting
> Thread interference detection

Figure 12.1 Evolution of PowerPC performance monitor unit.

function. The POWER5 brings some order to the situation by redefining the registers by function:

> **MMCR0**—Controls major features and when to count (freeze conditions, interrupt enabling, etc.)
> **MMCR1**—Controls what to count (event selection)
> **MMCRA**—Controls advanced features such as profiling support and thresholding

The POWER5 PMU is always enabled for counting (although by default the programmable counters count the null event, nothing). The counters can be programmed to count unconditionally or based on the state of the processor. In the PowerPC Architecture the *machine status register (MSR)* is used to describe and control the processor state. For example, when executing non-privileged applications, the processor is in *problem* state, designated by the problem state bit in the MSR. While executing privileged operating system code the processor is in *supervisor* state. The PMU can use the state of the processor to allow monitoring just the operating system or just an application program. The MSR has a special bit, the *performance monitor mark (PMM)*, to indicate that the current process has been identified for performance monitoring. The bit has no functional impact; it is only an indicator set by software. The PMU can be programmed to count only when the PMM bit is set or only when it is not set to isolate the performance of the monitored process. The *run latch* is another indicator bit similar to the PMM. Operating systems can use the run latch, a bit in the CTRL SPR, to indicate that they are executing a dispatchable task (i.e., not idle). Early machines used this bit to drive a *CPU busy* light on the machine's front panel. The PMU can be programmed to gate all events by the run latch, filtering out events occurring in the idle loop.

12.2.3 Event selection

The POWER5 PMU has three classes of events—bus, direct, and hybrid events—that are a combination of one or more event bus signals with special processing within the PMU. Direct events are events that are pervasive in nature or are so critical that they must be available regardless of the configuration of the event bus. Events such as processor cycles and instructions complete are examples of direct events.

Bus events are events that are routed to the PMU via a 32-bit event bus as shown in Figure 12.2. The event bus is logically divided into four 8-bit *byte lanes*. Bus events from the functional units are multiplexed down to three unit buses. Each byte lane can be configured from a different unit bus, but each unit bus can only be configured from a single unit (the unit buses are not byte selectable, when a unit is selected it selects the entire 32-bit bus).

Hybrid events combine one or more events with special processing within the PMU. The POWER5 is superscalar microprocessor with pairs of identical functional units (two fixed-point units, two floating-point units, two load/store units). To allow counting of individual unit events, each unit typically sends 4 bits of events per byte on the unit event bus. Figure 12.3 shows unit FPU0 using bits 0–3 and FPU1 using bits 4–7. This feature was first introduced on the POWER4. The PMU also has the ability to decode packed events (16 bits of information encoded into 4 bits of data and a valid bit, for example).

The processing of event signals (either discretely or via hybrid events) is designed generically. The PMU processes the signals the same way regardless of the source. This makes adding new events trivial. Discrete events can be placed anywhere on the event bus (although paired events must be placed identically within the upper and lower nibbles of a byte). Hybrid events only need to be placed appropriately on the event bus so that the PMU can process them.

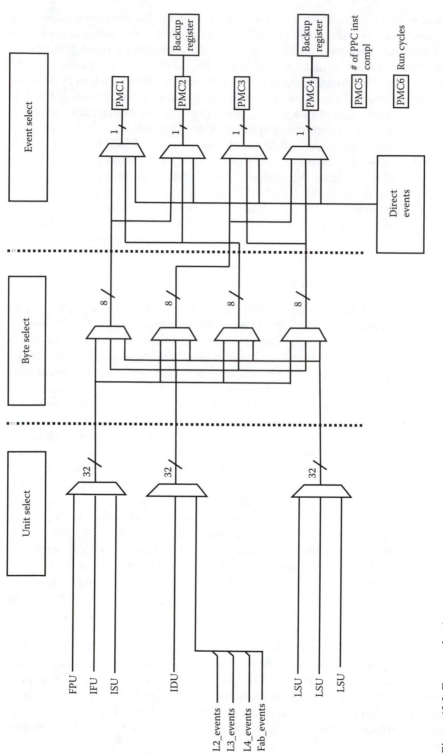

Figure 12.2 Event selection.

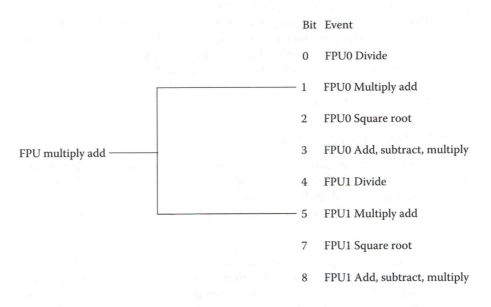

Bit	Event
0	FPU0 Divide
1	FPU0 Multiply add
2	FPU0 Square root
3	FPU0 Add, subtract, multiply
4	FPU1 Divide
5	FPU1 Multiply add
7	FPU1 Square root
8	FPU1 Add, subtract, multiply

FPU multiply add

Figure 12.3 Event layout for adding.

Simplicity in hardware implementation does have an effect on software support. Event selection is hierarchical in nature. With the exception of direct events, event selection first requires the configuration of the event bus. Each PMC is individually configured to select from a particular byte lane and then a bit within that byte lane. Some combinations of events are not possible (events from different units that are assigned to the same byte lane cannot be counted simultaneously).

12.3 Advanced features

12.3.1 Performance monitor exceptions

The PMU can be programmed to generate a *performance monitor exception* when particular conditions occur, such as a counter overflowing. When the processor is enabled for external interrupts a performance monitor exception will cause an interrupt. Because the exception is sticky, it persists until cleared, the PMU can be used to monitor kernel code where interrupts are disabled; any pending exception will cause an interrupt once they are enabled again. Once any performance monitor exception occurs, it disables future exceptions until the performance monitor interrupt handler enables them again. The POWER5 was designed to operate in a *logical partition (LPAR)* environment, where processors are shared by multiple partitions. To facilitate performance monitoring in a LPAR environment, the performance monitor exception is indicated by a MMCR bit. Performance monitor exceptions such as those caused by a counter overflow or other hardware condition set this bit automatically. Software can also set the performance monitor exception bit. If a partition being swapped out has a performance monitor exception outstanding but has not yet reflected the interrupt (because interrupts are disabled, for example), the pending exception can be preserved across a partition swap by saving and restoring the MMCRs and PMCs. The exception will eventually be reflected when the partition is dispatched again, and interrupts are enabled.

12.3.2   Data source feedback

Introduced with POWER4, *data source feedback* identifies where in the storage hierarchy an instruction fetch or data load was sourced from. Previous PowerPC processors had limited ability to identify L2 misses. The PPC604 used *load thresholding* to build a histogram of load resolution times [4]. This was compared to the known latency of the storage subsystem to infer the percentages of loads from each level. With the multilevel cache structure used on the POWER4, this technique was limiting (the latencies of some cache levels could not be differentiated from another). Data source feedback tags each load with an indicator of where it came from in the hierarchy. The PMU uses its decode capability to break the source tag into individual events for data sourced from a particular level. Instead of counting cache misses from each level, the POWER4 counts cache hits for both instructions and data. To determine cache misses at a given level, say, L2, would require that all of the hits from lower levels be counted and summed. POWER5 enhanced this technique by inspecting the load source and determining whether it was beyond the L2 and counting that as a L2 miss. It is important to note that the number of lines loaded into the L1 may not be equal to the number of L1 misses. The POWER5 has the ability to merge multiple misses to the same cache line, reducing the number of reloads.

12.3.3   Profiling support (sampling)

Sampling is commonly used to profile executing programs. Instruction sampling is the technique of identifying a particular instruction and collecting detailed information about that instruction. The information should be detailed enough to identify the instruction and condition(s) it encountered during execution. The PowerPC architecture provides two SPRs to identify sampled instructions. The *sampled instruction address register (SIAR)* captures the effective address of the sampled instruction. The *sampled data address register (SDAR)* captures the effective address of the sampled instruction's data operand, if any. An indicator bit in MMCRA indicates when the SIAR and SDAR are from the same instruction. The SDAR is not cleared when a new sampled instruction is selected, so the indicator bit is needed to show that the SDAR is not for a previous (since executed or cancelled) sampled instruction. The sampled registers can only be updated by the processor when performance monitor exceptions are enabled. Performance monitor exceptions toggle the enable bit, locking the contents of the sampled registers. This makes it possible to profile code that is not enabled for interrupts. Because the exception is persistent, it will eventually trigger an interrupt. The validity of the execution profile is influenced by the sampling technique. Some sampling techniques include

Event-based sampling—Sampling an instruction based on the occurrence of an event, such as a cache miss

Continuous sampling—Sampling every instruction executed

Slot-based sampling—Sample instructions based on some uniform criteria such as position in an internal queue

Random sampling—Sample instructions randomly

Each of these methods have benefits and drawbacks, particularly in a speculative execution, out-of-order processor [5]. The POWER5 processor can have over 100 instructions in its pipeline at any given time. Multiple storage accesses can be concurrently active, including more than one for the same cache line. Associating any monitored event with a particular instruction is challenging, and doing so without significantly increasing the overall area of the processor or impacting the critical pipeline stages is particularly difficult.

12.3.3.1   Historical background

The PowerPC 604 sampled based on the *load queue*. Instructions occupying the first slot in the load queue were sampled, or *marked*. Events caused by this instruction were annotated as marked events. Only load events could be marked, but on a RISC machine 20% or more of the instructions are loads and they heavily influence processor performance, making this was a reasonable approach. The PPC604 can have multiple loads in the load queue, each of them causing events that could cause a performance monitor exception. Profiling on marked events avoids this problem by ensuring that the sampled instruction was the source of the event. [6].

POWER3 continued the slot-based approach but used the reorder buffer (completion table). Sampling is no longer limited to load events. Slot-based sampling can induce bias in the sampled events because some instructions or sequences of instructions are more likely to end up in the instrumented slot than others. The POWER3 is a speculative execution processor, so some of the sampled instructions may never complete. It also executes out-of-order. Like the PPC604, the POWER3 uses marked events to identify events caused by the sampled instruction.

RS64 uses continuous sampling. The SIAR is updated on every completion, and the SDAR is updated for every load or store. Execution is in order and the processor has a short pipeline so the sampled instruction will be close to the event causing the performance monitor exception for most events and precise for some. There is no concept of a marked instruction or marked event.

To eliminate the problems caused by slot-based sampling, POWER4 introduced random sampling. The instruction decode unit (IDU) randomly picks an instruction from all *eligible* instructions. Eligibility can be PowerPC instructions, load/store instructions, instructions with a particular characteristic, or instructions that match a particular pattern (opcode matching). Because the POWER4 dispatches and tracks instructions in groups of up to five instructions, the captured instruction address is actually the address of the first instruction in the group. This is sufficient to profile down to the basic block level but is not sufficient to identify the particular instruction within the group that was sampled. As with the PPC604 and POWER3, marked events are used to identify events caused by marked instructions. Marked events are provided for performance-sensitive conditions, such as loads from a particular level of the storage system, translation misses, or flushes due to unaligned data. The effectiveness of random sampling is dependent on the sampling rate, how frequently an instruction is selected. Up to 20 dispatch groups can be active at once, and only one sampled instruction can be active at any given time, so the worst case sampling rate is 1 out of 20 groups. Assuming that each group has five instructions, the worst-case instruction sampling rate would be 1 out of 100. In reality the average group size is closer to four, even in perfectly scheduled code, bringing the worst-case sampling rate to 1 out of 80 instructions. Actual sampling rates vary from 1 out of 35 instructions (2.8% of all instructions) to 1 out of 50 (2% of all instructions). This sampling rate is close to the POWER3 (1 out of 32 instructions) without the potential bias of slot-based schemes.

12.3.3.2   POWER5 sampling

The POWER5 offers both random sampling and continuous sampling. With continuous sampling, every group that completes updates the SIAR and every load that causes a DL1 reload updates the SDAR. Continuous sampling is ideal for hot-spot analysis (either execution or data). Although continuous sampling provides better coverage for execution profiling (every group is sampled), it is limited to completion and data cache reload events. To profile on other events (such as the exact level in the memory system a reload was sourced from) random sampling is still useful. To provide better resolution with random sampling, POWER5 captures the position of the marked instruction within the dispatch group.

12.3.4   CPI breakdown

Cycles per instruction (CPI, or the reciprocal, instructions per cycle) is the fundamental metric in processor performance. Knowing the number of cycles (or fractions of a cycle) it takes to complete an instruction is critical to evaluating the performance of the processor. Knowing the components of CPI is also important (where did the cycles go?). CPI analysis is complicated on the POWER5 processor because it executes instructions out-of-order and completes instructions in groups. With up to five instructions per group, any one of the instructions could block completion. The POWER5, like the POWER4, is capable of completing one group per cycle (or up to five instructions per cycle). The POWER4 could count instructions completed, groups completed, and cycles to be used in CPI analysis. From these, CPI and cycles per group could be computed (as well as instructions per group to give insight into how well instructions are being scheduled by the compiler). It would be trivial to implement a completion stall event to count the cycles when no group completed, but without details on *why* nothing completed, the event would provide no added value. The same metric (cycles when nothing completed) can be easily computed by subtracting the number of groups completed (each group takes one cycle to complete) from total cycles. Complicating the analysis, the cause of the stall may not be known until after it clears. POWER5 aids the analysis of completion stalls by implementing speculative, or rewind, counters. Two of the programmable counters have backup registers that are not accessible to software. When software writes an initial value to the counter, its backup register is also written. A counter can be configured for a particular stall condition and begins counting on any cycle when no group is completing. The first group that completes will report the last condition that held its completion. If the condition matches what the counter is configured for, the count value is committed by updating the backup register. If it does not match, the counter is rewound to its previously checkpointed value. Table 12.1 shows the breakdown of CPI components and the events that used to calculate the breakdown. The shaded cells are measured directly.

Table 12.1 *Calculating CPI Breakdown*

Total cycle <# cycles>	Completion cycles <A:group complete cycles>	PPC Base completion cycles <A1: One or more PowerPC instructions completed this cycle>		
		overhead of cracking/microcoding <A2:(A)-(A1)>		
	Completion Table empty <B>	I-cache miss penalty <B1>		
		Branch redirection (branch misprediction) penalty <B2> PMC4SEL=0x38		
		others (Flush penalty etc) <B4: (B)-(B1)-(B2)>		
	Completion Stall cycles <C: total-(A)-(B)>	Stall by LSU inst <C1>	Stall by reject <C1A>	Stall by Translation (rejected by ERAT miss) <C1A1>
				other reject <C1A2: (C1A)-(C1A1)>
			Stall by D-cache miss <C1B>	
			Stall by LSU basic latency, LSU Flush penalty <C1C: (C1)-(C1A)-(C1B)>	
		Stall by FXU inst <C2>	Stall by any form of DIV/MTSPR/MFSPR inst <C2A>	
			Stall by FXU basic latency <C2C: (C2)-(C2A)>	
		Stall by FPU inst <C3>	Stall by any form of FDIV/FSQRT inst <C3A>	
			Stall by FPU basic latency <C3B: (C3)-(C3A)>	
		others (Stall by BRU/CRU inst, flush penalty (except LSU flush), etc) <C4: (completion stall cycles)-(C1)-(C2)-(C3) >		

The events are hierarchical. Load/store unit (LSU) stalls will include D-cache miss and rejects. Reject stalls will include Translation stalls due to *effective to real address translation (ERAT)* miss. This method only reports the last condition to clear and will not detect dependency chains. For example, a fixed-point *add* instruction that is dependent on a load in the same group that misses the D-cache will cause the completion stall condition to be reported as fixed-point unit (FXU). When multiple conditions are reported at the same time, load/store conditions are favored over all others, and D-cache miss is favored over rejects.

Completion can also be stalled because there are no instructions in the pipeline. The primary causes of this are instruction cache misses and branch mispredicts. The POWER5 has speculative events for these conditions that are handled in the same way as completion stalls.

12.3.5   Internal processor performance

The POWER5 uses SMT to maximize the performance of the processor. It does this by sharing resources between two threads to minimize idle resources. The POWER5 PMU monitors the processor to aid in the analysis of available resources. Functional units report instructions finished on a per-unit basis, which can be reported individually or summed by like units; this is useful to detect compiler scheduling problems that favor one unit over another. All of the critical queues within the processor are instrumented to report full conditions. Many queues, such as the completion table, are also instrumented to report empty conditions. The completion table and the instruction fetch buffer (instructions waiting for decode) are further instrumented to report on the number of entries used for each thread. The load/ store request queues and the load miss queue instrument the first queue slot to report the total number of allocations for that slot and the number of cycles when it was occupied. From this can be calculated the average queue occupancy and an estimate of fullness. For load misses it gives an estimate of average load latency for L1 misses. In addition to average load latency, the POWER5 PMU can capture cycles waiting on a particular level of the storage subsystem using the rewind counters. A counter is configured to count latency at a particular level, memory, for example. When a sampled load misses the L1, the latency counter begins counting. When the sampled load returns, its source information is checked against the level being counted. If it is correct, the counter is committed. If it is not correct, the counter is rewound to the previous value. Average storage system latency can be calculating by dividing the total count of cycles waiting for marked loads from a particular level by the number of marked loads from that level of the storage system.

These internal processor events are useful for both single-thread and multithread mode. The POWER5 PMU also collects information on SMT operation. Both threads reports the number of cycles it was running at each thread priority level (1 to 7) as well as the number of cycles at a priority level higher or lower than the other thread (e.g., my thread 1 higher than other thread). If my thread is preempted by the other thread, there are events available to my PMU to record the occurrences and their reasons.

12.4 Software support

12.4.1 AIX™

AIX 5.2 [7] introduced support for the PMAPI package [8]. The *libpmapi* library contains a set of application programming interfaces (APIs) that are designed to provide access to some of the counting facilities of the performance monitor feature included in selected IBM microprocessors. Those APIs include the following:

- A set of system-level APIs to allow counting of the activity of a whole machine or of a set of processes with a common ancestor
- A set of first-party, kernel-thread-level APIs to allow threads running in 1:1 mode to count their own activity
- A set of third-party, kernel-thread-level APIs to allow a debug program to count the activity of target threads running in 1:1 mode

System-level monitoring requires root access and requires exclusive use of the PMU facilities on all processors; only one process can use the system-level APIs at a time, and it prevents use of thread-level APIs. PMAPI virtualizes the 32-bit counters into 64-bit counters, handling the overflow detection. It also abstracts and virtualizes the MMCRs. Virtual control structures and counters can exist for each thread using the thread-level APIs; each thread can configure the counters independently.

As mentioned earlier, configuring the POWER5 PMU is a complex task. To reduce the complexity, PMAPI provides a set of predefined groups of events. The group concept was first seen on the RS64, which actually implemented event groups in hardware. It was extended to software managed groups for POWER4 to ease the complexity of configuration. A robust set of groups is defined that cover the most common uses, and new groups can be added. Figure 12.4 shows an example of typical predefined groups.

Calls to the PMAPI library can be inserted into program source code to allow monitoring of specific sections. Figure 12.5 shows the calls required to configure, start, stop, and collect counter data for a short section of code. For many studies it is more convenient to measure the entire application's execution. A sample program, *tcount*, is provided that will execute a specified command while collecting counter data. Figure 12.6 shows the output from tcount. If the section of code to be studied is small or can easily be isolated in the source code, calling the PMAPI library as shown in Figure 12.5 will produce the most accurate results. For measuring the entire application execution or if the section to be monitored is a significant portion of the

```
Group #0: pm_utilization
Group name: CPI and utilization data
Group description: CPI and utilization data
Group status: Verified
Group members:
Counter  1, event 190: PM_RUN_CYC   : Run cycles
Counter  2, event 71: PM_INST_CMPL  : Instructions completed
Counter  3, event 56: PM_INST_DISP  : Instructions dispatched
Counter  4, event 12: PM_CYC [shared] : Processor cycles
Counter  5, event  0: PM_INST_CMPL  : Instructions completed
Counter  6, event  0: PM_RUN_CYC   : Run cycles


Group #1: pm_completion
Group name: Completion and cycle counts
Group description: Completion and cycle counts
Group status: Verified
Group members:
Counter  1, event  2: PM_1PLUS_PPC_CMPL  : One or more PPC instruction
             completed
Counter  2, event 195: PM_GCT_EMPTY_CYC [shared] : Cycles GCT empty
Counter  3, event 49: PM_GRP_CMPL  : Group completed
Counter  4, event 12: PM_CYC [shared] : Processor cycles
Counter  5, event  0: PM_INST_CMPL  : Instructions completed
Counter  6, event  0: PM_RUN_CYC   : Run cycles


Group #2: pm_group_dispatch
Group name: Group dispatch events
Group description: Group dispatch events
Group status: Verified
Group members:
Counter  1, event 66: PM_GRP_DISP_VALID   : Group dispatch valid
Counter  2, event 65: PM_GRP_DISP_REJECT  : Group dispatch rejected
Counter  3, event 50: PM_GRP_DISP_BLK_SB_CYC  : Cycles group dispatch
             blocked by scoreboard
Counter  4, event 60: PM_INST_DISP  : Instructions dispatched
Counter  5, event  0: PM_INST_CMPL  : Instructions completed
Counter  6, event  0: PM_RUN_CYC   : Run cycles


Group #40: pm_branch_miss
Group name: Branch mispredict
Group description: Branch mispredict
Group status: Verified
Group members:
Counter  1, event 210: PM_TLB_MISS   : TLB misses
Counter  2, event 184: PM_SLB_MISS   : SLB misses
Counter  3, event  1: PM_BR_MPRED_CR  : Branch mispredictions due to CR
             bit setting
Counter  4, event  3: PM_BR_MPRED_TA  : Branch mispredictions due to
             target address
Counter  5, event  0: PM_INST_CMPL  : Instructions completed
Counter  6, event  0: PM_RUN_CYC   : Run cycles


Group #41: pm_branch1
Group name: Branch operations

Group description: Branch operations
Group status: Verified
Group members:
```

(a)

Figure 12.4 Sample counter groups.

```
Counter   1, event   9: PM_BR_UNCOND   : Unconditional branch
Counter   2, event   8: PM_BR_PRED_TA   : A conditional branch was
            predicted, target prediction
Counter   3, event   3: PM_BR_PRED_CR   : A conditional branch was
            predicted, CR prediction
Counter   4, event   5: PM_BR_PRED_CR_TA   : A conditional branch was
            predicted, CR and target prediction
Counter   5, event   0: PM_INST_CMPL   : Instructions completed
Counter   6, event   0: PM_RUN_CYC   : Run cycles


Group #42: pm_branch2
Group name: Branch operations
Group description: Branch operations
Group status: Verified
Group members:
Counter   1, event 64: PM_GRP_BR_REDIR_NONSPEC   : Group experienced non-
            speculative branch redirect
Counter   2, event 62: PM_GRP_BR_REDIR   : Group experienced branch
            redirect
Counter   3, event 24: PM_FLUSH_BR_MPRED   : Flush caused by branch
            mispredict
Counter   4, event 59: PM_INST_CMPL   : Instructions completed
Counter   5, event   0: PM_INST_CMPL   : Instructions completed
Counter   6, event   0: PM_RUN_CYC   : Run cycles


Group #43: pm_L1_tlbmiss
Group name: L1 load and TLB misses
Group description: L1 load and TLB misses
Group status: Verified
Group members:
Counter   1, event 20: PM_DATA_TABLEWALK_CYC   : Cycles doing data
            tablewalks
Counter   2, event 21: PM_DTLB_MISS   : Data TLB misses
Counter   3, event 100: PM_LD_MISS_L1   : L1 D cache load misses
Counter   4, event 106: PM_LD_REF_L1   : L1 D cache load references
Counter   5, event   0: PM_INST_CMPL   : Instructions completed
Counter   6, event   0: PM_RUN_CYC   : Run cycles


Group #44: pm_L1_DERAT_miss
Group name: L1 store and DERAT misses
Group description: L1 store and DERAT misses
Group status: Verified
Group members:
Counter   1, event 13: PM_DATA_FROM_L2   : Data loaded from L2
Counter   2, event 137: PM_LSU_DERAT_MISS   : DERAT misses
Counter   3, event 165: PM_ST_REF_L1   : L1 D cache store references
Counter   4, event 171: PM_ST_MISS_L1   : L1 D cache store misses
Counter   5, event   0: PM_INST_CMPL   : Instructions completed
Counter   6, event   0: PM_RUN_CYC   : Run cycles
```

(b)

Figure 12.4 (Continued).

```
Group #41: pm_branch1
Group name: Branch operations
Group description: Branch operations
Group status: Verified
Group members:
Counter  1, event   9: PM_BR_UNCOND   : Unconditional branch
Counter  2, event   8: PM_BR_PRED_TA  : A conditional branch was
             predicted, target prediction
Counter  3, event   3: PM_BR_PRED_CR  : A conditional branch was
             predicted, CR prediction
Counter  4, event   5: PM_BR_PRED_CR_TA  : A conditional branch was
             predicted, CR and target prediction
Counter  5, event   0: PM_INST_CMPL   : Instructions completed
Counter  6, event   0: PM_RUN_CYC     : Run cycles


Group #42: pm_branch2
Group name: Branch operations
Group description: Branch operations
Group status: Verified
Group members:
Counter  1, event 64: PM_GRP_BR_REDIR_NONSPEC  : Group experienced non-
             speculative branch redirect
Counter  2, event 62: PM_GRP_BR_REDIR  : Group experienced branch
             redirect
Counter  3, event 24: PM_FLUSH_BR_MPRED  : Flush caused by branch
             mispredict
Counter  4, event 59: PM_INST_CMPL  : Instructions completed
Counter  5, event  0: PM_INST_CMPL  : Instructions completed
Counter  6, event  0: PM_RUN_CYC    : Run cycles


Group #43: pm_L1_tlbmiss
Group name: L1 load and TLB misses
Group description: L1 load and TLB misses
Group status: Verified
Group members:
Counter  1, event 20: PM_DATA_TABLEWALK_CYC  : Cycles doing data
             tablewalks
Counter  2, event 21: PM_DTLB_MISS  : Data TLB misses
Counter  3, event 100: PM_LD_MISS_L1  : L1 D cache load misses
Counter  4, event 106: PM_LD_REF_L1  : L1 D cache load references
Counter  5, event  0: PM_INST_CMPL  : Instructions completed
Counter  6, event  0: PM_RUN_CYC    : Run cycles
```
(c)

Figure 12.4 (Continued).

overall executed instructions, a tool like tcount is more convenient and should be sufficiently accurate.

12.4.2   Linux

Several efforts are underway to add support of the POWER5 PMU to LINUX. SUSE®; LINUX Enterprise Server 9 has rudimentary support through the sysfs file system. The MMCRs and PMCs are exposed as files under /sys/

```
int which_group = 1;

grp_ptr=getenv("PM_GROUP");              /* allow PM_GROUP environmental
                                            variable */
if (grp_ptr != NULL) which_group = atoi(grp_ptr);


setprog.events[0] = which_group;         /* event group id            */
setprog.mode.b.is_group = 1;             /* POWER5 requires groups */
filter |= PM_GET_GROUPS;                 /* to return list of groups  */
filter |= PM_VERIFIED;                   /* want all events           */
filter |= PM_UNVERIFIED;
filter |= PM_CAVEAT;
setprog.mode.b.kernel=1;                 /* count in kernel mode      */
setprog.mode.b.user=1;                   /* count in user mode        */

if ( (rc = pm_initialize(filter, &Myinfo, &My_group_info,PM_CURRENT)) !=
OK_CODE) {
        pm_error("pm_initialize", rc);
        exit(ERROR_CODE);
    }
if ( (rc = pm_set_program_mythread(&setprog)) != OK_CODE) {
        pm_error("pm_set_program_mythread", rc);
        exit(ERROR_CODE);
    }
print_group(which_group);            /* procedure to print group */

if ( (rc = pm_start_mythread()) != OK_CODE) /* start counting!     */
        pm_error("pm_start", rc);

/* code to be measured goes here ***********************************/
    for (z=0;z<n;z+=4) {
        x[z] = y[z];
        x[z+1] = y[z+1];
        x[z+2] = y[z+2];
        x[z+3] = y[z+3];
        }
/* code to be measured goes here ***********************************/

if ( (rc = pm_stop_mythread()) != OK_CODE) /* stop counting!     */
        pm_error("pm_stop_mythread", rc);

if ( (rc = pm_get_program_mythread(&getprog)) != OK_CODE)
        pm_error("pm_get_program_mythread", rc);
if ( (rc = pm_get_data_mythread(&mydata)) != OK_CODE)
        pm_error("pm_get_data_mythread", rc);
                                    (a)
```

Figure 12.5 Sample use of AIX PMAPI calls.

264 *Performance Evaluation and Benchmarking*

```
/* print_data*/
for(j=0;j<Myinfo.maxpmcs;j++)
  printf("%-18lld ",data->accu[j]);
printf("\n");
```

(b)

Figure 12.5 (Continued).

```
/usr/pmapi/samples/tcount -pur -g 131 mytest 100000

called with 2 arguments, first one is 100000
loop count is 100000, array pad is 0
loop count is 100000
*** Configuration :
Mode = user; Process tree = off; Thresholding = off
Event Group is specified.
Group 131: pm_Dmiss
Counter   1, event  16: PM_DATA_FROM_L3
Counter   2, event  18: PM_DATA_FROM_LMEM
Counter   3, event 100: PM_LD_MISS_L1
Counter   4, event 171: PM_ST_MISS_L1
Counter   5, event   0: PM_INST_CMPL
Counter   6, event   0: PM_RUN_CYC

*** Results :
CPU      PMC 1          PMC 2          PMC 3          PMC 4          PMC 5          PMC 6
====  ============   ============   ============   ============   ============   ============
[ 0]  0              1670           10247          200569         2645054        3089454
[ 1]  0              0              0              0              0              0
[ 2]  0              0              0              0              8              0
[ 3]  0              0              0              0              8              0
====  ============   ============   ============   ============   ============   ============
ALL   0              1670           10247          200569         2645070        3089454
```

Figure 12.6 Sample tcount output.

devices/system/cpu/cpu*n*/. MMCRs and PMCs can be opened, read, and written using simple file operations.

12.5 Challenges

The POWER5 is an aggressively out-of-order, superscalar, and speculative processor capable of processing two independent streams of instructions per CPU, four per chip. This presents challenges to the design of performance instrumentation and the interpretation of data collected. The profiling support already described was designed to allow the performance analyst to identify particular instructions that encounter or cause performance sensitive conditions. The ability to combine data from pairs of execution units into a single count was designed to aggregate the data into more management metrics. Speculative execution presents a particularly difficult challenge. To track events only for instructions that complete would require a significant amount of logic and buffers to capture and maintain event data for all active instructions. Instead, the POWER5 counts most events as they happen (speculative or not). Analyzing this data requires some understanding of the speculative nature of the machine. Counts taken when they occur can be used with other such counts but could be misleading when used with counts taken at completion. Calculating the ratio of load instructions executed to all instructions executed is meaningful, but the ratio of load instructions executed to all instructions completed may not be. Some events, such as cache misses, are only available at execution. Calculating cache misses per completed instruction is still useful as a relative metric (comparing one workload to another on the same system, or one tuning option to another) but should not be considered an absolute measurement.

12.6 Summary

The POWER5 Microprocessor is an advanced modern processor. It can process two independent threads of execution simultaneously and dynamically balance resources between the two threads. To monitor and manage the performance of the processor requires advanced performance-monitoring techniques. The POWER5 performance monitor unit (PMU) is the culmination of evolutionary advances beginning with the PowerPC 604. This chapter only scratches the surface of the capability and use of the PMU. With full support for the hardware performance monitor standard in AIX, it is expected that its use will become widespread.

References

1. Sprunt, B., The basics of performance-monitoring hardware, *Micro, IEEE* 22, 4, 65–71, 2002.
2. Simultaneous multihreading (SMT) on eServer iSeries POWER5 processors, online at: http://www-1.ibm.com/servers/eserver/iseries/perfmgmt/pdf/SMT.pdf.
3. Roth, C., Levine, F., PowerPC performance monitor evolution, *Proceedings of IPCCC 1997*, 331–336. Feb. 1997.
4. Welbon, E.H., Moore, R.S., Levine, F.E., Roth, C.P., Load miss performance analysis methodology using the PowerPC 604 performance monitor for OLTP workloads, *Proceedings of Compcon '96*, 111–116, Feb. 1996.
5. Dean, J., Hicks, J, Waldspurger, C., Weihl, W., Chrysos, G., et al., ProfileMe: Hardware Support for Instruction-Level Profiling on Out-of-Order Processors, *Proceedings of Micro-30*, 292–302, December 1997.
6. Roth, C., Levine, F., Welbon, E., Performance monitoring on the PowerPC 604 microprocessor, *Proceedings of ICCD '95*, 212–215, Oct. 1995.
7. AIX 5L Version 5.2 Performance Management Guide (SC23-4876-00) International Business Machines Corporation 1997, 2004.
8. AIX 5L Version 5.2 Performance Tools Guide and Reference (SC23-4859-02) International Business Machines Corporation 1997, 2004.

Trademarks

The following terms are trademarks or registered trademarks of the IBM Corporation in the United States or other countries or both: PowerPC, POWER4, POWER5, PowerPC 970, PowerPC 604, PowerPC 604e, AIX, eServer, iSeries SUSE is a registered trademark of SUSE AG, a Novell business. Linux is a registered trademark of Linus Torvalds.

Chapter Thirteen

Performance Monitoring on the Itanium®* Processor Family

Rumi Zahir, Kishore Menezes, and Susith Fernando

Contents

* Registered trademark of Intel Corporation

267

13.1   Introduction

The Itanium architecture relies on the extraction of instruction-level parallelism (ILP) in software. One philosophy of the architecture is to enable the compiler to expose the parallelism in programs, thereby simplifying the hardware implementations. The architecture provides many features that the compiler can employ in accomplishing this task. The Itanium architecture [1] provides support for control and data speculation, allowing the compiler to reduce the impact of memory latency by breaking control and memory dependence barriers. Full predication support is also available that allows removal of branches to transform a control dependence to a data dependence. Such architectural support comes with the challenge to use these features judiciously. The Itanium architecture makes available a full-featured and extensive performance-monitoring unit (PMU) to ease the burden of addressing this challenge.

Performance analysis of workloads plays a very important part in microarchitecture tuning. It is essential to understand the execution properties of the workloads expected to be executed on the microarchitecture. The Itanium PMU allows for monitoring dynamic processor behavior. Information from this monitoring process can then be used to understand the behavior of a workload and to characterize it. The microarchitectural events captured by the performance monitoring unit help understand the effect of compiler optimizations on the workload, the use of architectural features such as speculation and predication, and the effectiveness of microarchitectural structures such as the advanced load address table (ALAT), the caches and the translation lookaside buffers (TLBs). In addition to microarchitectural improvements, these measurements provide the data to drive application tuning and future processor, compiler, and operating system designs.

Another application of performance monitoring is to understand the execution characteristics of a program on a given machine. Profile information [2] can be used to optimize the execution of the profiled program. Some of the earliest work on profiling concentrated on obtaining information about the execution times of the functions within a given program. The same could also be applied to algorithms. The statistics gathered were then used to optimize the functions with the highest execution times or to make improvements to badly implemented algorithms.

The scope and applications of profiling have change over the years. Profile information is steadily gaining importance in the optimization of program execution, especially in the area of hand tuning of programs [2], trace scheduling [3], superblock scheduling [4], data preloading [5], branch prediction [6], and improved instruction cache performance [7]. Issues in the exploitation of instruction-level parallelism inherent in programs, coupled with rapid developments in compiler research, have generated interest in the role of profile information in smart compilation.

Traditionally, profiling and the use of profiles have been achieved through a long, tedious instrument-run-recompile sequence. During instrumentation, the compiler inserts additional instructions into the original program to collect accurate execution frequencies of basic blocks or the arcs that connect these basic blocks. Next, this instrumented code is executed with a variety of batch inputs. After these multiple executions, the program statistics are calculated based on the profile information that has been collected. Finally, the original program is recompiled using profile-driven optimizations. Not only do these traditional techniques suffer from the need for multiple execution and compilation passes, they also suffer from slowdown of the program being profiled due to the added instructions. The MIPS basic block profiling tool *pixie* [8] inserts about five instructions in every basic block [9]. Ball and Larus measured the slowdown of pixie required for arc-based profiling as between 1.11 to 5.24 times [9]. To offset this slowdown, modifications involving the reduction in the number of blocks or the arcs that need to be probed have been suggested in the literature. Ball and Larus investigated one such technique that reduces the slowdown to a maximum of 2.05 times for the SPEC92 benchmarks. Unfortunately, this overhead is still too large for software vendors to readily absorb. Commercial software vendors can tolerate only a negligible amount of additional execution overhead (< = 5%) [10].

If profiling is to gain commercial acceptance, it must be smoothly integrated into the software development cycle. Unfortunately, this requires the reduction or elimination of the need for a sample input suite, as well as more efficient profiling methods. The use of performance monitors integrated into the processor allows software vendors to analyze and optimize applications with no profile collection overhead. The advantage to using performance monitors is twofold: It eliminates the need for sample input suites, and the optimizations are based on actual program usage. Profiling in this manner is commercially appealing because vendors' alpha and beta testing processes are often very well-defined and the hardware-based style of profiling leverages their existing investment to produce better-optimized code.

This chapter defines the performance monitoring features of the Itanium processor family. The Itanium 2* processor provides four 48-bit performance counters, more than 100 events that can be monitored, and several advanced monitoring capabilities. This chapter outlines the targeted performance monitor usage models and defines the software interface and programming model.

13.2 Workload characterization and microarchitecture tuning

The first step in any performance analysis is to understand the performance characteristics of the workload under study. There are two fundamental measures of interest: event rates and program cycle break down.

---

* Registered trademark of Intel Corporation.

13.2.1 Event rate monitoring

Event rates of interest include average retired instructions per cycle (IPC), data and instruction cache miss rates, or branch misprediction rates measured across the entire application. Characterization of operating systems or large commercial workloads (e.g., online transaction processing) requires a system-level view of performance relevant events such as TLB miss rates, virtual hash page table (VHPT) walks per second, interrupts per second, or bus utilization rates. Event rate monitoring determines event rates by reading the processor event occurrence counters before and after the workload is run, and then computing the desired rates. For instance, two basic Itanium processor events that count the number of retired Itanium instructions (IA64_INST_RETIRED) and the number of elapsed clock cycles (CPU_CYCLES) allow a workload's IPC to be computed as follows:

$$IPC = (IA64\_INST\_RETIRED_{t1} - IA64\_INST\_RETIRED_{t0}) / (CPU\_CYCLES_{t1} - CPU\_CYCLES_{t0})$$

Time-based sampling is the basis for many performance debugging tools [2,12,13]. Time-based sampling can be used to plot the event rates over time, and can provide insights into the different phases of the workload.

On the Itanium processor, many event types (e.g., TLB misses or branch mispredictions) are limited to a rate of one per clock cycle. These are referred to as *single occurrence* events. However, multiple events of the same type may occur in the same clock. Such events are referred to as *multi-occurrence* events. An example of a multi-occurrence event on the Itanium processor is data cache read misses. There can be up to two data cache misses per clock. Multi-occurrence events, such as the number of entries in the memory request queue, can be used to derive the average number and average latency of memory accesses. The next two subsections describe the basic Itanium processor mechanisms for monitoring single- and multi-occurrence events.

13.2.1.1 Single-occurrence events and duration counts

A single-occurrence event can be monitored by any of the Itanium processor performance counters with a few exceptions. For all single-occurrence events, a counter is incremented by up to one per clock cycle. Duration counters that count the number of clock cycles during which a condition persists are considered single-occurrence events. Examples of single-occurrence events on the Itanium processor are TLB misses, branch mispredictions, and cycle-based metrics.

13.2.1.2 Multi-occurrence events, thresholding, and averaging

Events that, due to hardware parallelism, may occur at rates greater than one per clock cycle are termed multi-occurrence events. Examples of such events on the Itanium processor are retired instructions or the number of live entries in the memory request queue.

When dealing with multi-occurrence events, it is sometimes useful to have the ability to count when the number of events of a certain type exceed a certain threshold. Thresholding capabilities are available in the Itanium processor's multi-occurrence counters and can be used to plot an event distribution histogram. When a nonzero threshold is specified, the monitor is incremented by one in every cycle in which the observed event count exceeds that programmed threshold. This capability allows microarchitectural buffer sizing experiments to be supported by real measurements. For example, measurements could help find the number of cycles during which the memory request queue contained more than two entries or the number of cycles during which more than three instructions were retired. By running a benchmark with different threshold values, a histogram can be drawn up that may help to identify the performance "knee" at a certain buffer size.

For overlapping concurrent events, such as pending memory operations, the average number of concurrently outstanding requests, and the average number of cycles that requests were pending are of interest. To calculate the average number or latency of multiple outstanding requests in the memory queue, we need to know the total number of requests (n_{total}) and the number of live requests per cycle (n_{live}/cycle). By summing up the live requests (n_{live}/cycle) using a multi-occurrence counter, Σn_{live} is directly measured by hardware. We can now calculate the average number of requests and the average latency as follows:

$$\text{Average outstanding requests/cycle} = \Sigma n_{live}/\Delta t$$
$$\text{Average latency per request} = \Sigma n_{live}/n_{total}$$

An example of this calculation is given in Table 13.1. The average outstanding requests per cycle = 15/8 = 1.825, and the average latency per request = 15/5 = 3 cycles.

The Itanium processor provides the following capabilities to support event rate monitoring:

- Clock cycle counter
- Retired instruction counter
- Event occurrence and duration counters
- Multi-occurrence counters with thresholding capability

Table 13.1 Example of Average Latency per Request and Requests per Cycle Calculation

Time [Cycles]	1	2	3	4	5	6	7	8
# Requests In	1	1	1	1	1	0	0	0
# Requests Out	0	0	0	1	1	1	1	1
n_{live}	1	2	3	3	3	2	1	0
n_{live}	1	3	6	9	12	14	15	15
n_{total}	1	2	3	4	5	5	5	5

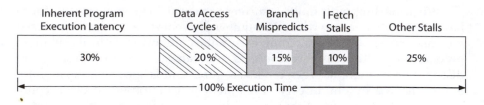

Figure 13.1 Itanium® processor family cycle accounting.

13.2.2 *Cycle accounting*

Although event rate monitoring counts the number of events, it does not tell us whether the observed events contribute to a performance problem. A commonly used strategy is to plot multiple event rates and correlate them with the measured IPC rate. If a low IPC occurs concurrently with a peak of cache miss activity, chances are that cache misses are causing a performance problem. To eliminate such guesswork, the Itanium processor provides a set of cycle accounting monitors that break down the number of cycles that are lost due to various kinds of microarchitectural events. As shown in Figure 13.1, this lets us account for every cycle spent by a program and therefore provides insight into an application's microarchitectural behavior. Note that cycle accounting is different from simple stall or flush duration counting. Cycle accounting is based on the machine's actual stall and flush conditions, and accounts for overlapped pipeline delays, whereas simple stall or flush duration counters do not. Cycle accounting determines a program's cycle breakdown by stall and flush reasons, whereas simple duration counters are useful in determining cumulative stall or flush latencies.

The Itanium processor cycle accounting monitors account for all major single- and multi-cycle stall and flush conditions. Overlapping stall and flush conditions are prioritized in reverse pipeline order (i.e., delays that occur later in the pipe and that overlap with earlier stage delays are reported as being caused later in the pipeline). The six back-end stall and flush reasons are prioritized in the following order:

1. Exception/interruption cycle: Cycles spent flushing the pipe due to interrupts and exceptions
2. Branch misprediction cycle: Cycles spent flushing the pipe due to branch mispredictions
3. Data/FPU access cycle: Memory pipeline full, data TLB stalls, load-use stalls, and access to floating-point unit
4. Execution latency cycle: Scoreboard and other register dependency stalls
5. RSE active cycle: RSE spill/fill stall
6. Front-end stalls: Stalls due to the back end waiting on the front end

Additional front-end stall counters are available, which detail seven possible reasons for a front-end stall to occur. However, the back-end and front-end stall events should not be compared, because they are counted in different stages of the pipeline.

13.3   Profiling

Profiling is used by application developers, profile-guided compilers, optimizing linkers, and run-time systems. Application developers are interested in identifying performance bottlenecks and relating them back to their source code. Based on profile feedback, developers can make changes to the high-level algorithms and data structures of the program. Compilers can use profile feedback to optimize instruction schedules by employing advanced Itanium architectural features such as predication and speculation.

To support profiling, performance monitor counts have to be associated with program locations. The following mechanisms are supported directly by the Itanium processor's performance monitors:

- Program counter sampling
- Miss event address sampling: Itanium processor event address registers (EARs) provide subpipeline-length event resolution for performance-critical events (instruction and data caches, branch mispredictions, and instruction and data TLBs).

These profiling features are presented in the next two subsections.

13.3.1   Program counter sampling

Application tuning tools [2,12] use time-based or event-based sampling of the program counter and other event counters to identify performance critical functions and basic blocks. The sampled points can be represented in a histogram by instruction addresses. For application tuning, statistical sampling techniques have been very successful because the programmer can rapidly identify code hot spots in which the program spends a significant fraction of its time, or where certain event counts are high.

Program counter sampling can point a performance analyst at code hot spots, but it does not indicate the cause of the performance problem. Inspection and manual analysis of the hot-spot region along with a fair amount of guesswork are required to identify the root cause of the performance problem. On the Itanium processor, the cycle accounting mechanism described in subsection 13.2.2 can be used to directly measure an application's microarchitectural behavior.

The Itanium architectural interval timer facilities (ITC and ITM registers) can be used for time-based program counter sampling. Event-based program counter sampling is supported by a dedicated performance monitor overflow interrupt.

To support program counter sampling, the Itanium processor provides the following mechanisms:

- Timer interrupt for time-based program counter sampling
- Event count overflow interrupt for event-based program counter sampling
- Hardware-supported cycle accounting

13.3.2 *Miss event address sampling*

Program counter sampling and cycle accounting provide an accurate picture of cumulative microarchitectural behavior, but they do not provide the application developer with pointers to specific program elements (code locations and data structures) that repeatedly cause microarchitectural miss events. In a cache study of the SPEC92 benchmarks, Lebeck used trace based cache miss profiling to gain performance improvements of 1.02 to 3.46 on various benchmarks by making simple changes to the source code [11]. This type of analysis requires identification of instruction and data addresses related to microarchitectural miss events such as cache misses, branch mispredictions, or TLB misses. Using symbol tables or compiler annotations, these addresses can be mapped back to critical source code elements. Like Lebeck, most performance analysts in the past have had to capture hardware traces and resort to trace driven simulation.

Due to the superscalar issue, deep pipelining, and out-of-order instruction completion of today's microarchitectures, the sampled program counter value may not be related to the instruction address that caused a miss event. On a Pentium processor pipeline, the sampled program counter may be off by two dynamic instructions from the instruction that caused the miss event. On a Pentium Pro processor, this distance increases to approximately 32 dynamic instructions. On the Itanium processor, it is approximately 48 dynamic instructions. If program counter sampling is used for miss-event address identification on the Itanium processor, a miss event might be associated with an instruction almost five dynamic basic blocks away from where it actually occurred (assuming that 10% of all instructions are branches). Therefore, it is essential for hardware to precisely identify an event's address.

The Itanium processor provides a set of *event address registers (EARs)* that record the instruction and data addresses of data cache misses for loads, the instruction and data addresses of data TLB misses, and the instruction addresses of instruction TLB and cache misses. A four-entry-deep branch trace buffer captures sequences of branch instructions. Table 13.2 summarizes the capabilities offered by the Itanium processor EARs and the branch trace buffer. Exposing miss-event addresses to software allows them to be monitored either by sampling or by code instrumentation. This eliminates the need for trace generation to identify and solve performance problems and enables performance analysis by a much larger audience on unmodified hardware.

Table 13.2 Itanium Processor EARs and Branch Trace Buffer

Event Address Register	Triggers On	What Is Recorded
Instruction Cache	Instruction fetches that miss the L1 instruction cache (demand fetches only)	Instruction address Number of cycles fetch was in flight
Instruction TLB (ITLB)	Instruction fetch missed L1 ITLB (demand fetches only)	Instruction address Who serviced L1 ITLB miss: L2 ITLB, VHPT, or software
Data cache	Load instructions that miss L1 data cache	Instruction address Data address Number of cycles load was in flight.
Data TLB (DTLB)	Data references that miss L1 DTLB	Instruction address Data address Who serviced L1 DTLB miss: L2 DTLB, VHPT, or software
Branch trace buffer	Branch outcomes	Branch instruction address Branch target instruction address Mispredict status and reason

The Itanium processor EARs enable statistical sampling by configuring a performance counter to count, for instance, the number of data cache misses or retired instructions. The performance counter value is set up to interrupt the processor after a predetermined number of events have been observed. The data cache event address register repeatedly captures the instruction and data addresses of actual data cache-load misses. Whenever the counter over-flows, miss-event address collection is suspended until the event address register is read by software (this prevents software from capturing a miss event that might be caused by the monitoring software itself). When the counter overflows, an interrupt is delivered to software, the observed event addresses are collected, and a new observation interval can be setup by rewriting the performance counter register. For time-based (rather than event-based) sampling methods, the event address registers indicate to software whether a qualified event was captured. Statistical sampling can achieve arbitrary event resolution by varying the number of events within an observation interval and by increasing the number of observation intervals.

13.4 Event qualification

Many of the performance-monitoring events on the Itanium processor can be qualified in a number of ways such that only a subset of the events are counted using performance-monitoring counters. As shown in Figure 13.2,

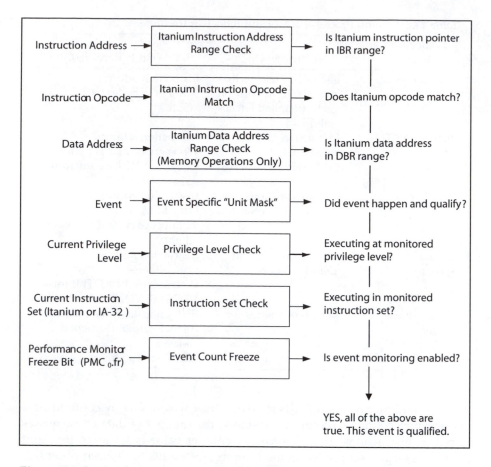

Figure 13.2 Itanium processor event qualification.

events can be qualified for monitoring based on instruction address range, instruction opcode, data address range, event-specific unit mask (umask), the privilege level and instruction set the event was caused by, and the status of the performance-monitoring freeze bit ($PMC_0.fr$). The following list describes these capabilities in detail.

- Itanium instruction address range check: The Itanium processor allows event monitoring to be constrained to a programmable instruction address range. This enables monitoring of dynamically linked libraries (DLLs), functions, or loops of interest in the context of a large Itanium-based application. The Itanium instruction address range check is applied at the instruction fetch stage of the pipeline, and the resulting qualification is carried by the instruction throughout the pipeline. This enables conditional event counting at a level of granularity smaller than dynamic instruction length of the pipeline (approximately 48 instructions).

- Itanium instruction opcode match: The Itanium processor provides two independent Itanium opcode match registers, each of which match the currently issued instruction encodings with a programmable opcode match and mask function. The resulting match events can be selected as an event type for counting by the performance counters. This enables creating histograms of instruction types, usage of destination, and predicate registers as well as basic block profiling (through insertion of tagged NOPs).
- Itanium data address range check: The Itanium processor allows event collection for memory operations to be constrained to a programmable data address range. This enables selective monitoring of data-cache miss behavior of specific data structures.
- Event-specific unit masks: Some events allow the specification of unit masks to filter out interesting events directly at the monitored unit. As an example, the number of counted bus transactions can be qualified by an event specific unit mask to contain transactions that originated from any bus agent, from the processor itself, or from other input/output (I/O) bus masters. In this case, the bus unit uses a three-way unit mask (any, self, or I/O) that specifies which transactions are to be counted. In the Itanium processor, events from the branch, memory, and bus units support a variety of unit masks.
- Privilege level: Two bits in the processor status register (PSR) are provided to enable selective process-based event monitoring. The Itanium processor supports conditional event counting based on the current privilege level, which allows performance-monitoring software to break down event counts into user and operating system contributions.
- Instruction set: The Itanium processor supports conditional event counting based on the currently executing instruction set (Itanium or IA-32) by providing two instruction-set mask bits for each event monitor. This allows performance-monitoring software to break down event counts into Itanium and IA-32 contributions.
- Performance monitor freeze: Event counter overflows or software can freeze event monitoring. When frozen, no event monitoring takes place until software clears the monitoring freeze bit ($PMC_0.fr$). This ensures that the performance-monitoring routines themselves (e.g., counter overflow interrupt handlers or performance monitoring context switch routines) do not pollute the event counts of the system under observation.

13.4.1 Combining opcode matching, instruction, and data address range check

The Itanium processor allows various event qualification mechanisms to be combined by providing the instruction tagging mechanism shown in Figure 13.3.

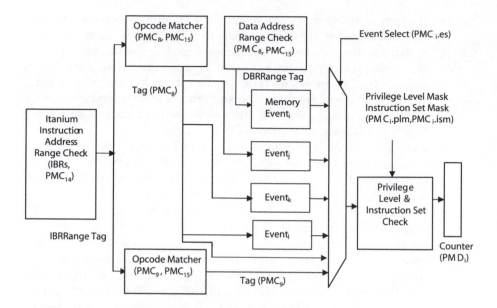

Figure 13.3 Instruction tagging mechanism in the Itanium processor.

During Itanium instruction execution, the instruction address range check is applied first. The resulting instruction address range check tag (IBRRangeTag) is passed to two opcode matchers that combine the instruction address range check with the opcode match. Each of the two combined tags can be counted as a retired instruction count event.

One of the combined Itanium address range and opcode match tags, Tag(PMC$_8$), qualifies all downstream pipeline events. Events in the memory hierarchy (L1 and L2 data cache and data TLB events) can further be qualified using a data address range check tag (DBRRangeTag).

As summarized in Table 13.3, data address range checking can be combined with opcode matching and instruction-range checking on the Itanium processor. Additional event qualifications based on the current privilege level can be applied to all events and are discussed in subsection 13.4.2.

13.4.2 Privilege level constraints

Performance-monitoring software cannot always count on context switch support from the operating system. In general, this has made performance analysis of a single process in a multiprocessing system or a multiprocess workload impossible. To provide hardware support for this kind of analysis, the Itanium architecture specifies three global bits—user performance monitor enable (PSR.up), privileged performance monitor enable (PSR.pp), and interruption privileged performance monitor enable (DCR.pp)—and a per-monitor privilege monitor bit (PMC$_i$.pm). To break down the performance

Table 13.3 Itanium Processor Event Qualification Modes

Event Qualification Modes	Opcode Match Enable	Opcode Matching PMC_8	Instruction Address Range Check Enable	Data Address Range Check (Mem Pipe Events Only)
Unconstrained monitoring (all events)	x	0xffff_ffff _ffff_ffff	x	[1,11] or [0,xx]
Instruction address range check only; channel 0	x	0xffff_ffff _ffff_fffe	0	[1,00]
Opcode matching only	1	Desired Opcodes	x	[1,01]
Data address range check only	x	0xffff_ffff _ffff_ffff	x	[1,10]
Instruction address range check and opcode matching, channel0	1	Desired Opcodes	0	[1,01]
Instruction and data address range check	x	0xffff_ffff _ffff_fffe	0	[1,00]
Opcode matching and data address range check	1	Desired Opcodes	x	[1,00]

contributions of operating system and user-level application components, each monitor specifies a 4-bit privilege level mask (PMC_i.plm). The mask is compared to the current privilege level in the processor status register (PSR.cpl), and event counting is enabled if PMC_i.plm[PSR.cpl] is 1.

PMC registers can be configured as user-level monitors (PMC_i.pm is 0) or system-level monitors (PMC_i.pm is 1). A user-level monitor is enabled whenever PSR.up is 1. PSR.up can be controlled by an application using the set user mask (*sum*) and reset user mask (*rum*) instructions. This allows applications to enable/disable performance monitoring for specific code sections. A system-level monitor is enabled whenever PSR.pp is 1. PSR.pp can be controlled at privilege level 0 only, which allows monitor control without interference from user-level processes. The pp field in the default control register (DCR.pp) is copied into PSR.pp whenever an interruption is

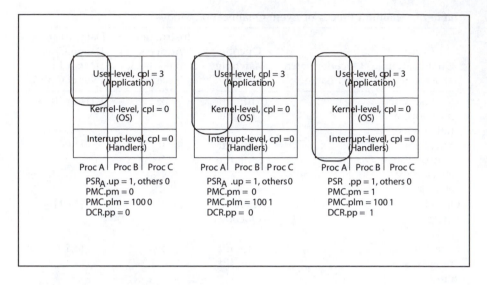

Figure 13.4 Single-process monitor.

delivered. This allows events generated during interruptions to be broken down separately: If DCR.pp is 0, events during interruptions are not counted; if DCR.pp is 1, they are included in the kernel counts.

As shown in Figures 13.4, 13.5, and 13.6, single-process, multi-process, and system-level performance monitoring are possible by specifying the appropriate combination of PSR and DCR bits. These bits allow

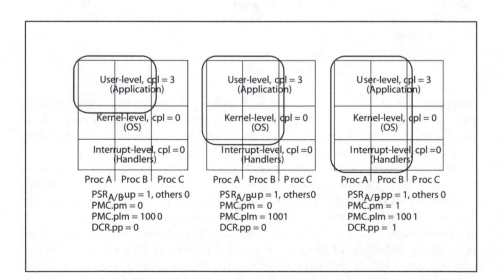

Figure 13.5 Multiple-process monitor.

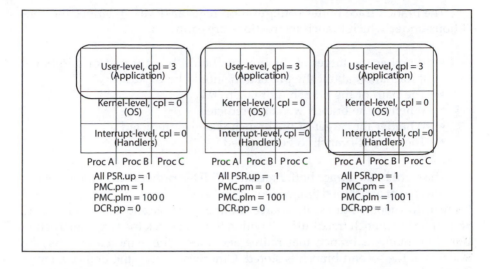

Figure 13.6 System-wide monitor.

performance monitoring to be controlled entirely from a kernel-level device driver, without explicit operating system support. Once the desired monitoring configuration has been setup in a process's processor status register (PSR), "regular" unmodified operating context switch code automatically enables/disables performance monitoring.

With support from the operating system, individual per-process break-down of event counts can be generated.

13.5   Branch trace buffer

The branch trace buffer on the Itanium 2 processor provides information about the outcome of the most recent Itanium branch instructions and their predictions and outcomes. The Itanium 2 branch trace buffer configuration register (PMC_{12}) defines the conditions under which branch instructions are captured and allows the trace buffer to capture specific subsets of branch events.

In every cycle in which a qualified Itanium branch retires, its source bundle address and slot number are written to the branch trace buffer. The branch's target address is written to the next buffer location. If the target instruction bundle itself contains a qualified Itanium branch, the branch trace buffer either records a single trace buffer entry (with *b*-bit set) or makes two trace buffer entries: one that records the target instruction as a branch target (*b*-bit cleared) and another that records the target instruction as a branch source (*b*-bit set). As a result, the branch trace buffer may contain a mixed sequence of branches and targets.

The branch trace buffer configuration register (PMC_{12}) defines the conditions under which branch instructions are captured:

- Whether the target of the branch should be captured or additional information about the prediction should be captured
- The path of the branch (not taken/taken)
- Whether the branch was mispredicted
- Whether the target of the branch was mispredicted
- The type of branch to be captured

The eight branch trace buffer registers PMD_{8-15} provide information about the outcome of a captured branch sequence. In every cycle in which a qualified branch instruction retires, its source bundle address and slot number are written to the branch trace buffer. If within the next clock the target instruction bundle contains a branch that retires and meets the same conditions, the address of the second branch is stored. Otherwise, either the branch's target address or details of the branch prediction are written to the next buffer location.

The Itanium 2 branch trace buffer is a circular buffer containing the last four to eight qualified Itanium branches. The branch trace buffer index register (PMD_{16}) identifies the most recently recorded branch or target. In every cycle in which a qualified branch or target is recorded, the branch buffer index is postincremented. After eight entries have been recorded, the branch index wraps around, and the next qualified branch will overwrite the first trace buffer entry. The wrap condition itself is also recorded by means of a bit in PMD_{16}.

13.6 Summary

The Itanium architecture and its microarchitectural implementations incorporate extensive performance-monitoring capabilities. Innovative performance characterization features such as cycle accounting and event rate monitoring ease the task of performance analysis. This chapter has provided an overview of these and other performance analysis capabilities. For more details the reader is referred to *The Intel Itanium 2 Processor Reference Manual* [14].

Acknowledgments

The authors would like to thank Jeremy Williamson and Allan Knies of Intel Corporation for their contributions to this chapter.

References

1. *IA-64 Application Developer's Architecture Guide.* Intel Corporation, online at: http://developer.intel.com/design/ia64/devinfo.htm, 1999.
2. Graham S.L., Kessler, P.B., and McKusick, M.K., gprof: A call graph execution profiler, *Proceedings SIGPLAN'82 Symposium on Compiler Construction; SIGPLAN Notices* 17, 6, 120–126, 1982.
3. Fisher, J.A., Trace scheduling: A technique for global microcode compaction, *IEEE Trans. Computers* C-30, 7, 478–490, July 1981.
4. Chang, P.P., Mahlke, S.A., and Hwu, W.W., Using profile information to assist classic code optimization, *Software—Practice and Experience* 21, 1301-1321, Dec. 1991.
5. Chen, W.Y., *Data preload for superscalar and VLIW processors.* Ph.D. thesis, Dept. of Electrical and Computer Engineering, University of Illinois, Urbana-Champaign, 1993.
6. McFarling, S., and Hennessy, J.L., Reducing the cost of branches, in *Proc. 13th Ann. Int'l Symp. Computer Architecture* (Tokyo, Japan), 396–403, June 1986.
7. Hwu, W.W. and Chang, P.P., Achieving high instruction cache performance with an optimizing compiler, in *Proc. 16th Ann. Int'l. Symp. Computer Architecture* (Jerusalem, Israel), 242–251, May 1989.
8. MIPS Computer Systems, *UMIPS-V Reference Manual*, Sunnyvale, CA, 1990.
9. Ball, T. and Larus, J.R., Optimally profiling and tracing programs, in *Proc. of the ACM SIGPLAN 1992 Conference on Principles of Programming Languages*, 59–70, 1992.
10. Cox, J.S., Howell, D.P., and Conte, T.M., Commercializing profile-driven optimization, in Proc. 28th Hawaii Int'l. Conf. On System Sciences (Maui, HI), 1, 221–228, Jan. 1995.
11. Lebeck, Alvin R., and Wood, David A., Cache profiling and the SPEC benchmarks: A case study, Tech Report 1164, Computer Science Dept., University of Wisconsin–Madison, July 1993.
12. Atkins, Mark, and Subramaniam, Ramesh, PC software performance tuning, *IEEE Computer*, 29, 8, 47–54, 1996.
13. Blake, Russ, Optimizing Windows NT(tm), *Microsoft "Windows NT Resource Kit for Windows NT Version 3.51,"* 4, Microsoft Press, 1995.
14. *Intel Itanium 2 Processor Reference Manual.* Intel Corporation, online at: ftp://download.intel.com/design/Itanium2/manuals/25111003.pdf, May 2004.

Index